Quantifying Matter

Joseph A. Angelo, Jr.

Facts On File
An Infobase Learning Company

This book is dedicated to the memory of my beloved daughter,
Jennifer April Angelo (1975–1993)

Facts On File, Inc.
An imprint of Infobase Learning
132 West 31st Street
New York NY 10001

Library of Congress Cataloging-in-Publication Data
Angelo, Joseph A.
    Quantifying matter / Joseph A. Angelo, Jr.
       p. cm.—(States of matter)
    Includes bibliographical references and index.
    ISBN 978-0-8160-7609-3
    1. Matter—Constitution—Juvenile literature. I. Title.
    QC173.16.A54 2011
    530—dc22   2010015829

Facts On File books are available at special discounts when purchased in bulk quantities for businesses, associations, institutions, or sales promotions. Please call our Special Sales Department in New York at (212) 967-8800 or (800) 322-8755.

You can find Facts On File on the World Wide Web at http://www.factsonfile.com

Excerpts included herewith have been reprinted by permission of the copyright holders; the author has made every effort to contact copyright holders. The publishers will be glad to rectify, in future editions, any errors or omissions brought to their notice.

Text design and composition by Annie O'Donnell
Illustrations by Sholto Ainslie
Photo research by the Author
Cover printed by Bang Printing, Inc., Brainerd, Minn.
Book printed and bound by Bang Printing, Inc., Brainerd, Minn.
Date printed: February 2011
Printed in the United States of America

10 9 8 7 6 5 4 3 2 1

This book is printed on acid-free paper.

Contents

Preface

The unleashed power of the atom has changed everything save our modes of thinking.

—Albert Einstein

Humankind's global civilization relies upon a family of advanced technologies that allow people to perform clever manipulations of matter and energy in a variety of interesting ways. Contemporary matter manipulations hold out the promise of a golden era for humankind—an era in which most people are free from the threat of such natural perils as thirst, starvation, and disease. But matter manipulations, if performed unwisely or improperly on a large scale, can also have an apocalyptic impact. History is filled with stories of ancient societies that collapsed because local material resources were overexploited or unwisely used. In the extreme, any similar follies by people on a global scale during this century could imperil not only the human species but all life on Earth.

Despite the importance of intelligent stewardship of Earth's resources, many people lack sufficient appreciation for how matter influences their daily lives. The overarching goal of States of Matter is to explain the important role matter plays throughout the entire domain of nature—both here on Earth and everywhere in the universe. The comprehensive multivolume set is designed to raise and answer intriguing questions and to help readers understand matter in all its interesting states and forms—from common to exotic, from abundant to scarce, from here on Earth to the fringes of the observable universe.

The subject of matter is filled with intriguing mysteries and paradoxes. Take two highly flammable gases, hydrogen (H_2) and oxygen (O_2), carefully combine them, add a spark, and suddenly an exothermic reaction takes place yielding not only energy but also an interesting new substance called water (H_2O). Water is an excellent substance to quench a fire, but it is also an incredibly intriguing material that is necessary for all life here on Earth—and probably elsewhere in the universe.

Matter is all around us and involves everything tangible a person sees, feels, and touches. The flow of water throughout Earth's biosphere, the air people breathe, and the ground they stand on are examples of the most commonly encountered states of matter. This daily personal encounter with matter in its liquid, gaseous, and solid states has intrigued human beings from the dawn of history. One early line of inquiry concerning the science of matter (that is, *matter science*) resulted in the classic earth, air, water, and fire elemental philosophy of the ancient Greeks. This early theory of matter trickled down through history and essentially ruled Western thought until the Scientific Revolution.

It was not until the late 16th century and the start of the Scientific Revolution that the true nature of matter and its relationship with energy began to emerge. People started to quantify the properties of matter and to discover a series of interesting relationships through carefully performed and well-documented experiments. Speculation, philosophical conjecture, and alchemy gave way to the scientific method, with its organized investigation of the material world and natural phenomena.

Collectively, the story of this magnificent intellectual unfolding represents one of the great cultural legacies in human history—comparable to the control of fire and the invention of the alphabet. The intellectual curiosity and hard work of the early scientists throughout the Scientific Revolution set the human race on a trajectory of discovery, a trajectory that not only enabled today's global civilization but also opened up the entire universe to understanding and exploration.

In a curious historical paradox, most early peoples, including the ancient Greeks, knew a number of fundamental facts about matter (in its solid, liquid, and gaseous states), but these same peoples generally made surprisingly little scientific progress toward unraveling matter's inner mysteries. The art of metallurgy, for example, was developed some 4,000 to 5,000 years ago on an essentially trial-and-error basis, thrusting early civilizations around the Mediterranean Sea into first the Bronze Age and later the Iron Age. Better weapons (such as metal swords and shields) were the primary social catalyst for technical progress, yet the periodic table of chemical elements (of which metals represent the majority of entries) was not envisioned until the 19th century.

Starting in the late 16th century, inquisitive individuals, such as the Italian scientist Galileo Galilei, performed careful observations and measurements to support more organized inquiries into the workings of the natural world. As a consequence of these observations and experiments, the nature of matter became better understood and better quantified. Scientists introduced the concepts of density, pressure, and temperature in their efforts to more consistently describe matter on a large (or macroscopic) scale. As instruments improved, scientists were able to make even better measurements, and soon matter became more clearly understood on both a macroscopic and microscopic scale. Starting in the 20th century, scientists began to observe and measure the long-hidden inner nature of matter on the atomic and subatomic scales.

Actually, intellectual inquiry into the microscopic nature of matter has its roots in ancient Greece. Not all ancient Greek philosophers were content with the prevailing earth-air-water-fire model of matter. About 450 B.C.E., a Greek philosopher named Leucippus and his more well-known student Democritus introduced the notion that all matter is actually composed of tiny solid particles, which are *atomos* (ατομος), or indivisible. Unfortunately, this brilliant insight into the natural order of things lay essentially unnoticed for centuries. In the early 1800s, a British schoolteacher named John Dalton began tinkering with mixtures of gases and made the daring assumption that a chemical element consisted of identical indestructible atoms. His efforts revived atomism. Several years later, the Italian scientist Amedeo Avogadro announced a remarkable hypothesis, a bold postulation that paved the way for the atomic theory of chemistry. Although this hypothesis was not widely accepted until the second half of the 19th century, it helped set the stage for the spectacular revolution in matter science that started as the 19th century rolled into the 20th.

What lay ahead was not just the development of an atomistic kinetic theory of matter, but the experimental discovery of electrons, radioactivity, the nuclear atom, protons, neutrons, and quarks. Not to be outdone by the nuclear scientists, who explored nature on the minutest scale, astrophysicists began describing exotic states of matter on the grandest of cosmic scales. The notion of degenerate matter appeared as well as the hypothesis that supermassive black holes lurked at the centers of most large galaxies after devouring the masses of millions of stars. Today, cosmologists and astrophysicists describe matter as being clumped into enormous clusters and superclusters of galaxies. The quest for these scientists is to explain how the observable universe, consisting of understandable forms of matter and energy, is also immersed in and influenced by mysterious forms of matter and energy, called dark matter and dark energy, respectively.

The study of matter stretches from prehistoric obsidian tools to contemporary research efforts in nanotechnology. States of Matter provides 9th- to 12th-grade audiences with an exciting and unparalleled adventure into the physical realm and applications of matter. This journey in search of the meaning of substance ranges from everyday "touch, feel, and see" items (such as steel, talc, concrete, water, and air) to the tiny, invisible atoms, molecules, and subatomic particles that govern the behavior and physical characteristics of every element, compound, and mixture, not only here on Earth, but everywhere in the universe.

Today, scientists recognize several other states of matter in addition to the solid, liquid, and gas states known to exist since ancient times. These include very hot plasmas and extremely cold Bose-Einstein condensates. Scientists also study very exotic forms of matter, such as liquid helium (which behaves as a superfluid does), superconductors, and quark-gluon plasmas. Astronomers and astrophysicists refer to degenerate matter when they discuss white dwarf stars and neutron stars. Other unusual forms of matter under investigation include antimatter and dark matter. Perhaps most challenging of all for scientists in this century is to grasp the true nature of dark energy and understand how it influences all matter in the universe. Using the national science education standards for 9th- to 12th-grade readers as an overarching guide, the States of Matter set provides a clear, carefully selected, well-integrated, and enjoyable treatment of these interesting concepts and topics.

The overall study of matter contains a significant amount of important scientific information that should attract a wide range of 9th- to 12th-grade readers. The broad subject of matter embraces essentially all fields of modern science and engineering, from aerodynamics and astronomy, to medicine and biology, to transportation and power generation, to the operation of Earth's amazing biosphere, to cosmology and the explosive start and evolution of the universe. Paying close attention to national science education standards and content guidelines, the author has prepared each book as a well-integrated, progressive treatment of one major aspect of this exciting and complex subject. Owing to the comprehensive coverage, full-color illustrations, and numerous informative sidebars, teachers will find the States of Matter to be of enormous value in supporting their science and mathematics curricula.

Specifically, States of Matter is a multivolume set that presents the discovery and use of matter and all its intriguing properties within the context of science as inquiry. For example, the reader will learn how the ideal gas law (sometimes called the ideal gas equation of state) did not happen overnight. Rather, it evolved slowly and was based on the inquisitiveness and careful observations of many scientists whose work spanned a period of about 100 years. Similarly, the ancient Greeks were puzzled by the electrostatic behavior of certain matter. However, it took several millennia until the quantified nature of electric charge was recognized. While Nobel Prize–winning British physicist Sir J. J. (Joseph John) Thomson was inquiring about the fundamental nature of electric charge in 1898, he discovered the first subatomic particle, which he called the *electron*. His work helped transform the understanding of matter and shaped the modern world. States of Matter contains numerous other examples of science as inquiry, examples strategically sprinkled throughout each volume to show how scientists used puzzling questions to guide their inquiries, design experiments, use available technology and mathematics to collect data, and then formulate hypotheses and models to explain these data.

States of Matter is a set that treats all aspects of physical science, including the structure of atoms, the structure and properties of matter, the nature of chemical reactions, the behavior of matter in motion and when forces are applied, the mass-energy conservation principle, the role of thermodynamic properties such as internal energy and entropy (disorder principle), and how matter and energy interact on various scales and levels in the physical universe.

The set also introduces readers to some of the more important solids in today's global civilization (such as carbon, concrete, coal, gold, copper, salt, aluminum, and iron). Likewise, important liquids (such as water, oil, blood, and milk) are treated. In addition to air (the most commonly encountered gas here on Earth), the reader will discover the unusual properties and interesting applications of other important gases, such as hydrogen, oxygen, carbon dioxide, nitrogen, xenon, krypton, and helium.

Each volume within the States of Matter set includes an index, an appendix with the latest version of the periodic table, a chronology of notable events, a glossary of significant terms and concepts, a helpful list of Internet resources, and an array of historical and current print sources for further research. Based on the current principles and standards in teaching mathematics and science, the States of Matter set is essential for readers who require information on all major topics in the science and application of matter.

Acknowledgments

I thank the public information and/or multimedia specialists at the U.S. Department of Energy (including those at DOE Headquarters and at all the national laboratories), the U.S. Department of Defense (including the individual armed services: U.S. Air Force, U.S. Army, U.S. Marines, and U.S. Navy), the National Institute of Standards and Technology (NIST) within the U.S. Department of Commerce, the U.S. Department of Agriculture, the National Aeronautics and Space Administration (NASA) (including its centers and astronomical observatory facilities), the National Oceanic and Atmospheric Administration (NOAA) of the U.S. Department of Commerce, and the U.S. Geological Survey (USGS) within the U.S. Department of the Interior for the generous supply of technical information and illustrations used in the preparation of this set of books. Also recognized here are the efforts of executive editor Frank K. Darmstadt and the other members of the Facts On File team, whose careful attention to detail helped transform an interesting concept into a polished, publishable product. The continued support of two other special people must be mentioned here. The first individual is my longtime personal physician, Dr. Charles S. Stewart III, whose medical skills allowed me to successfully work on this interesting project. The second individual is my wife, Joan, who for the past 45 years has provided the loving and supportive home environment so essential for the successful completion of any undertaking in life.

Introduction

The history of civilization is essentially the story of the human mind understanding matter. *Quantifying Matter* presents many of the most important intellectual achievements and technical developments that led to the scientific interpretation of substance. Readers will discover how the ability of human beings to relate the microscopic (atomic level) behavior of matter to readily observable macroscopic properties (such as density, pressure, and temperature) has transformed the world.

Quantifying Matter describes the basic characteristics and properties of matter. The three most familiar states of matter found on Earth's surface are solid, liquid, and gas. When temperatures get sufficiently high, another state of matter, called plasma, appears. Finally, as temperatures get very low and approach absolute zero, scientists encounter a fifth state of matter, called the Bose-Einstein condensate (BEC).

The rise of civilization parallels the ability of humans to measure and physically describe matter. The English word *quantitative* is an adjective that means "pertaining to or susceptible of measurement." When scientists quantify matter, they are attempting to describe certain physical properties or characteristics of matter in an orderly, understandable, and useful manner. Precise, repeatable measurements and experiments support the scientific understanding of matter. This book explains how scientists learned to measure matter and to quantify some of its most fascinating and useful properties.

Understanding the true nature of matter has been a very gradual process for most of human history. For example, the notion of the atom originated in ancient Greece about 2,400 years ago, yet the majority of the important concepts, breakthrough experiments, and technical activities surrounding the discovery and application of the atomic nucleus took place within the last 100 years or so.

The incredible story of matter starts with the beginning of the universe and an extremely energetic event, called the big bang. *Quantifying Matter* describes the cosmic origin of the elements. From prehistory to the present, the story of people parallels the human mind's restless search for the scientific meaning of substance. This quest began in earnest with the ancient Greek philosophers and continues today with modern scientists, who are exploring matter on its tiniest quantum levels as well as on its grandest cosmic scales. While science has unveiled some of matter's most intriguing mysteries, other aspects of matter and its intimate relationship with energy—such as dark matter and dark energy—remain elusive and unexplained.

Several thousand years ago, the use of metals and other materials supported the rise of sophisticated civilizations all around the Mediterranean Sea, with the Roman Empire serving as an enormous political stimulus and evolutionary endpoint. The Bronze Age and the Iron Age represent two important materials science–driven milestones in human development.

Before the Scientific Revolution, people in many civilizations practiced an early form of chemistry, known as alchemy. For most practitioners, alchemy was a mixture of mystical rituals, magical incantations, and random experimentation involving chemical reactions and procedures. Despite its often haphazard practices, medieval alchemy eventually gave rise to the science of chemistry in western Europe. As described in *Quantifying Matter,* early alchemists were influenced in their thinking by the four classic elements of ancient Greece, namely earth, air, water, and fire.

Despite the mysticism, rituals, and secrecy often associated with alchemy, the alchemists managed to contribute to the foundation of modern chemical science. Taken as a whole, they used trial and error techniques to explore all available materials and stumbled upon many interesting chemical reactions. The alchemists also developed several important techniques still used in modern chemistry, including distillation, fermentation, sublimation, and solution preparation.

As shown in *Quantifying Matter,* the initial breakthroughs in accurately quantifying matter took place during the Scientific Revolution. The volume highlights the pioneering efforts of several key individuals whose discoveries and insights established the foundations of physics, chemistry, and other physical sciences. The scientific method, developed as part of the Scientific Revolution, continues to exert a dominant influence on human civilization. Historically aligned with the Scientific Revolution was the development of the steam engine and the major technosocial transformation called the First Industrial Revolution.

This book also examines one of matter's most interesting and important properties—electromagnetism. Readers will learn how 18th-century scientists such as Benjamin Franklin began exploring the fundamental nature of electricity. At the dawn of the 19th century, the invention of the first electric battery (called Volta's electric pile) supported the emergence of the Second Industrial Revolution. Later in the 19th century, Scottish scientist James Clerk Maxwell presented a comprehensive set of equations that described electromagnetism. As the 20th century approached, Thomas Edison, Nikola Tesla, and other inventors harnessed electromagnetism in numerous innovative devices that gave humankind surprising new power and comforts.

The chemical properties of matter remained a mystery until woven into an elegant intellectual tapestry called the periodic table. *Quantifying Matter* describes the origin, content, and application of the periodic table. The reader will learn how Russian chemist Dmitri Mendeleev became the first scientist to propose a successful periodic table. In 1869, he suggested that when organized according to their atomic weights in a certain row-and-column manner, the known chemical elements exhibited a periodic behavior of their chemical and physical properties. This volume further describes how the current version of the periodic table was greatly expanded in content and meaning by the rise of modern atomic theory and quantum mechanics.

Readers will learn about the serendipitous discovery of radioactivity and how that fortuitous event led to the modern understanding of matter at the atomic and subatomic levels. *Quantifying Matter* discusses how scientists then went on to explore the atomic nucleus and learn how to harvest the incredible amounts of energy hidden within. The volume also describes how the modern understanding of matter on the smallest (quantum) scale continues in the 21st century, highlighted by the relentless quest for the Higgs boson, or God particle.

While the often bizarre behavior of matter at the atomic and subatomic levels defies human intuition and daily experience, *Quantifying Matter* explains how quantum mechanics greatly influences the lives of almost every person on Earth. The global information revolution depends on how devices such as transistors manipulate individual electric charges on the atomic scale. When people enjoy using their digital cameras, they are taking advantage of something called the photoelectric effect.

The book concludes with an introductory discussion of nanotechnology, the combined scientific and industrial effort involving the atomic level manipulation of matter, one atom or molecule at a time. Like many of the great milestones in materials science history, these ongoing research efforts promise to significantly alter the trajectory of civilization.

Quantifying Matter generates a comprehensive appreciation of how the principles and concepts of materials science govern many scientific fields, including astronomy, biology, medical sciences, chemistry, Earth science, meteorology, geology, mineralogy, oceanography, and physics. The scientific understanding of matter also forms an integral part of all branches of engineering, including aeronautics and astronautics, biological engineering, chemical engineering, civil engineering, environmental and sanitary engineering, industrial engineering, mechanical engineering, nuclear engineering, and ocean engineering.

The volume has been carefully designed to help any student or teacher who has an interest in the measurement or general behavior of matter discover what matter is, how scientists measure and characterize its various forms, and how the fascinating properties and characteristics of matter have influenced the course of human civilization. The back portion of the book contains an appendix with a contemporary periodic table, a chronology, a glossary, and an array of historical and current sources for further research. These should prove especially helpful for readers who need additional information on specific terms, topics, and events.

The author has carefully prepared *Quantifying Matter* so that a person familiar with SI units (International System of Units, the international language of science) will have no difficulty understanding and enjoying its contents. The author also recognizes that there is a continuing need for some students and teachers in the United States to have units expressed in United States customary units. Wherever appropriate, both unit systems appear side by side. An editorial decision places the American customary units first, followed by the equivalent SI units in parentheses. This format does not imply the author's preference for American customary units over SI units. Rather, the author strongly encourages all readers to take advantage of the particular formatting arrangement to learn more about the important role that SI units play within the international scientific community.

CHAPTER 1

Exploring the Nature of Matter

"Nothing exists except atoms and space . . ."
—Democritus, Greek philosopher (fifth century B.C.E.)

This chapter introduces some of the basic concepts about matter from the ancient Greek philosophers up to modern physicists and chemists. The nature and origin of matter has perplexed human beings from the dawn of time. While science has unveiled some of matter's most intriguing mysteries, other aspects remain mysterious and unexplained.

THE ATOMIC NATURE OF MATTER

Today, almost every student in high school or college has encountered the basic scientific theory that matter consists of atoms. But this widely accepted model of matter was not prevalent throughout most of human history.

Scientists trace the basic concept of the atomic structure of matter (that is, atomism) back to ancient Greece. In the fifth century B.C.E., the Greek philosopher Leucippus and his famous pupil Democritus (ca. 460–ca. 370 B.C.E.) introduced the theory of atomism within an ancient society that was quite comfortable assuming that four basic elements—earth, air, water, and fire—composed all the matter in the world. In founding the school of atomism, they speculated that matter actually consisted of an infinite number of minute, indivisible, solid particles.

Democritus pursued this interesting line of thought much further than his mentor. As a natural philosopher but not an experimenter, Democritus used his mind to consider what would happen if he kept cutting a piece of matter of any type into finer and finer halves. He reasoned that he would eventually reach a point where any further division would be impossible. He called these final indivisible pieces *atoma*. As a result, the modern word *atom* comes from the ancient Greek word *atomos* (ατομος), which means "not divisible."

For Democritus, these tiny, indivisible particles always existed and could never be destroyed. Different substances resulted from the way atoms connected, or linked together. Like the majority of the ancient Greek natural philosophers, his concepts resulted from hypothesis and the exercise of logic rather than rigorous experimentation and observation. (The scientific method involves theory, observation, and experimentation.) Despite the lack of experimentation, the ideas of Democritus concerning the innermost structure of matter were genuinely innovative some 2,500 years ago and represented the beginning of atomic theory.

Unfortunately for Democritus, the famous Greek philosopher Aristotle (384–22 B.C.E.) did not favor atomism, so the concept of atoms languished in the backwaters of Greek thinking. The innovative insights of another early Greek thinker, the astronomer Aristarchus (ca. 320–250 B.C.E.), suffered a similar fate due to the influence of Aristotle's teachings. Around 260 B.C.E., Aristarchus suggested that Earth was a planet that moved around the Sun. Aristotle had promoted geocentric cosmology—a cosmology that regarded Earth as stationary and the center of the universe. The sheer power of Aristotle's reputation for genius suppressed for centuries any support for Aristarchus's bold concept of heliocentric cosmology. Centuries passed before a Polish church official and astronomer, Nicholas Copernicus (1473–1543), revived the concept of heliocentric cosmology in 1543. Copernicus's actions started the great movement in Western civilization known as the Scientific Revolution. Powered by heliocentric (Sun-centered) cosmology, the Scientific Revolution swept away many of the Aristotelian concepts about nature that had impeded or restricted scientific progress for some 2,000 years.

During the 16th and 17th centuries, natural philosophers and scientists such as Galileo Galilei (1564–1642), René Descartes (1596–1650), Robert

Boyle (1627–91), and Sir Isaac Newton (1642–1727) all favored the view that matter was not continuous in nature but rather consisted of discrete, tiny particles, or atoms. However, it was not until the 19th century and the hard work of several pioneering chemists and physicists that the concept of the atom was gradually transformed from a vague philosophical concept into a modern scientific reality.

Science historians generally credit the British schoolteacher and chemist John Dalton (1766–1844) with the revival of atomism and the start of modern atomic theory. In 1803, Dalton suggested that each chemical element was composed of a particular type of atom. He defined the atom as the smallest particle, or unit of matter, in which a particular element can exist. His interest in the behavior of gases allowed Dalton to quantify the atomic concept of matter. He showed how the relative masses or weights of different atoms could be determined. To establish his relative scale, he assigned the atom of hydrogen a mass of unity. In so doing, Dalton revived atomic theory and inserted atomism back into the mainstream of scientific thinking.

Another important step in the emergence of atomic theory occurred in 1811, when the Italian scientist Amedeo Avogadro (1776–1856) formulated the famous hypothesis that eventually became known as Avogadro's law. He proposed that equal volumes of gases at the same temperature and pressure contain equal numbers of molecules. At the time, no one really had a clear understanding of the difference between an atom and a molecule. Later that century, scientists began to recognize the molecule as the smallest particle of any substance (element or compound) as it normally occurs in nature. In 1869, the Russian chemist Dmitri Ivanovich Mendeleev (1834–1904) published his famous periodic law. By the beginning of the 20th century, most scientists accepted the concept that molecules, such as water (H_2O), consisted of collections of atoms (here, two hydrogen atoms and one oxygen atom) and that atoms were the smallest units of matter with the properties of a chemical element.

Stimulated by Dalton's atomic hypothesis, other chemists busied themselves with identifying new elements and compounds. By 1819, the Swedish chemist Jöns Jacob Berzelius (1779–1848) had increased the number of known chemical elements to 50. In that year, Berzelius proposed the modern symbols for chemical elements and compounds based on abbreviations of the Latin names of the elements. He used the symbol O for oxygen (*oxygenium*), Cu for copper (*cuprum*), Au for gold (*aurum*), and so forth.

Over his lifetime, the brilliant chemist estimated the atomic weights of more than 45 elements, several of which he had personally discovered, including thorium (Th), which he identified in 1828.

While some 19th-century chemists filled in the periodic table, other scientists, such as the British physicist Michael Faraday (1791–1867) and the Scottish physicist James Clerk Maxwell (1831–79), explored matter's intriguing magnetic and electrical properties. Their pioneering work prepared the way for the German physicist Max Planck (1858–1947) to introduce his quantum theory in 1900 and for the German-Swiss-American physicist Albert Einstein (1879–1955) to introduce his theory of special relativity in 1905. Planck and Einstein provided the two great pillars of contemporary physics: quantum theory and relativity. Their great intellectual accomplishments formed the theoretical platform upon which other scientists in the 20th century constructed a more comprehensive theory of the atom, explored the intriguing realm of the atomic nucleus and the amazing world of subatomic particles and their energetic processes, exploited the equivalence of energy and matter through nuclear fission and fusion, and discovered the particle-wave duality of matter.

CARBON'S INTRIGUING PARADOX

This section presents one of the most interesting physical paradoxes found in the field of materials science. Why are a clear, sparkling diamond (the mineral that scratches glass) and a black-gray lump of graphite (the mineral that allows a pencil to leave a mark on paper) so different? This is a true mystery of nature because both objects consist of the very same substance, the chemical element carbon (C). Despite their common elemental makeup, the diamond is commonly regarded as the hardest available natural substance, while the mineral graphite is soft enough to be used as a lubricant or in a writing instrument. Furthermore, while diamond is a good electric insulator, graphite can conduct electricity. Scientists in previous centuries puzzled about these dramatic property differences of the same element without much success.

It was only in the 20th century that scientists eventually achieved a satisfactory explanation. The solution required an accurate physical understanding of what really lies at the heart of all matter—atoms. Scientists define an *atom* as the smallest particle of matter that retains its identity as a chemical element. Atoms are indivisible by chemical means and are the fundamental building blocks of all matter. The chemical

A brilliant cut 2.78-carat diamond   *(© 2002 Gemological Institute of America [GIA]; used with permission)*

elements, such as hydrogen (H), helium (He), carbon (C), iron (Fe), lead (Pb), and uranium (U), differ from one another because they consist of different types of atoms. According to (much simplified) modern atomic theory, an atom consists of a dense inner core (called the *nucleus*) that contains protons and neutrons surrounded by a distant cloud of orbiting electrons. When atoms are electrically neutral, the number of positively charged protons equals the number of negatively charged electrons. The number of protons in an atom's nucleus determines what chemical element it is, while how an atom shares its orbiting electrons determines the way that particular element physically behaves and chemically reacts. Through the phenomenon of covalent bonding, for example, an atom forms physically strong links by sharing one or more of its electrons with neighboring atoms. This atomic-scale linkage ultimately manifests itself in large-scale (that is, macroscopic) material properties, such as a substance's strength and hardness.

A sample of the mineral graphite from President Franklin Pierce's mine in New Hampshire *(Author)*

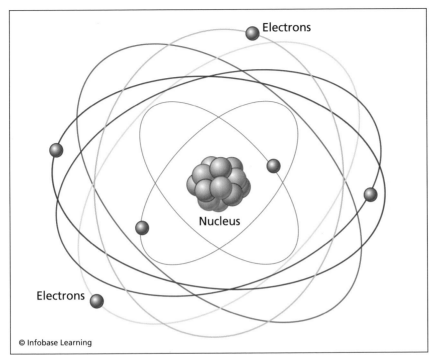

Simplified model of the carbon-12 atom showing the atom's six orbiting electrons, as well as the six protons (green) and six neutrons (red) in its nucleus. *(Illustration not drawn to scale) (DOE)*

THE DIFFERENCE BETWEEN PURE SUBSTANCES AND MIXTURES

Scientists find it convenient to treat matter as being either a *pure substance* or a *mixture*. A chemical element is a pure substance that consists of all the same type of atoms. The metals copper (Cu) and gold (Au) are familiar examples of pure substances. When viewed microscopically, these metallic elements consist entirely of the same type of atoms throughout. A pure substance can also consist entirely of identical molecules. Scientists define a *molecule* as a collection of atoms held together by chemical (bonding) forces. The atoms in a particular molecule may be identical, such as oxygen (O_2), or different, such as the compound molecules carbon monoxide (CO) and carbon dioxide (CO_2). A molecule is the smallest unit of matter that can exist by itself and still retain all the chemical properties of the element or compound.

Matter can also exist as a mixture of two or more substances in variable proportions. In a mixture, each substance retains its identity and does not change chemically. The air everyone breathes here on Earth, for example, is a mixture of several gases—predominantly nitrogen (N_2) molecules and oxygen (O_2) molecules—accompanied by much smaller quantities of other gases, including argon (Ar) and carbon dioxide (CO_2). Pouring salt (sodium chloride [NaCl]) molecules into water (H_2O) creates a homogenous mixture, which scientists call a saline solution. While the amount of salt dissolved in water can vary up to a saturation limit, the salt remains salt and the water remains water (from a chemical perspective) in this uniform-looking solution. In contrast, the olive oil, water, distilled vinegar, and variety of spices commonly found in Italian style salad dressings form a heterogeneous mixture—a mixture the appearance of which is not the same throughout. Vigorous shaking of this heterogeneous mixture is usually necessary to produce the quasi homogenous mixture of dressing that people like to pour on their salads.

Atoms are the basic building blocks of matter. As pure substances, the minerals diamond and graphite both contain only carbon atoms. However, the carbon atoms are arranged in significantly different ways, thereby creating different molecular forms called *allotropes*. The carbon atoms in diamond are held together by covalent bonds, which form a rigid three-dimensional crystal lattice that disperses light very well. Since every carbon atom in diamond is tightly bonded to four other carbon atoms, diamond has a rigid chain (or network) that results in the hardest commonly known natural mineral.

Graphite is another allotrope of carbon in which covalent bonds form sheets of atoms in hexagonal patterns. However, the adjacent sheets of carbon atoms in graphite are more loosely bound to each other by much weaker intermolecular forces, called van der Waals forces. The result of this particular molecular arrangement is startling, since graphite is also a crystalline mineral—but one that is very soft and brittle.

This familiar American coin contains an incredible number of copper atoms. *(U.S. Mint)*

This amazing colorized image was created by the custom-built scanning tunneling microscope at the National Institute of Standards and Technology (NIST) as scientists dragged a cobalt atom across a closely-packed lattice of copper atoms. *(Joseph Stroscio; Robert Celotta/NIST)*

Physicists like to characterize matter as anything that occupies space and has mass. Mass involves the interesting physical phenomenon of inertia, or a material object's resistance to any change in its state of motion. Since electrically neutral matter is self-attractive, matter gives rise to the very important force in nature called gravity (gravitation). Here on the surface of Earth, people usually describe how much mass an object has by expressing its weight. This is not scientifically correct, because the weight of an object is really a force—the local acceleration of gravity acting upon a mass.

Confusing? Perhaps. But this is why scientists needed to establish a set of rules when they attempted to quantify matter and energy during the Scientific Revolution. These rules are called scientific laws, statements about the physical world governed by widely accepted principles supported by many repeatable experiments and a self-consistent system of measurements.

Before scientists could intelligently quantify and accurately describe the many intriguing physical and chemical properties that matter exhibits both here on Earth and elsewhere in the universe, they needed to create a generally agreed-upon system of units. Today, the expressions *half a mole of a substance, five kilograms of mass,* and *one newton of force* all have specific, technical meanings to scientists around the world. The International System of Units, almost universally abbreviated SI, is a marvelous construct of the human mind and a superb example of global cooperation in science and engineering. Having an internationally agreed-upon system of units for physical measurements accelerated the rate of scientific progress and permitted unprecedented levels of cooperation on complex technical projects. But this was not always the case. For most of human history, the process of acquiring knowledge about the nature of matter was very gradual and filled with misconceptions, intellectual blind alleys, and sharp disagreements.

SCIENTIFIC UNITS

Science is based upon repeatable experiments, accurate measurements, and a system of logical, standardized units. The International System of Units is the generally agreed-upon, coherent system of units now in use throughout the world for scientific, engineering, and commercial purposes. Universally abbreviated SI (from the French Système International d'Unités), this system is also called the modern metric system.

The contemporary SI system traces its origins back to France in the late 18th century. During the French Revolution, government officials decided to introduce a decimal measurement system. Science historians view the deposit of two platinum standards (representing the meter and the kilogram) on June 22, 1799, in the Archives de la République as the first step in the development of the modern SI system. As a complement to these platinum standards for length and mass, scientists used astronomical techniques to precisely define a second of time.

Consequently, the fundamental SI units for length, mass, and time—the meter (m) [British spelling: *metre*], the kilogram (kg), and the second (s)—were based on natural standards. Modern SI units still rely on natural standards and international agreement, but these standards are now ones that can be measured with much greater precision than the previously used natural standards. Originally, scientists defined the meter as one-ten-millionth the distance from the equator to the North Pole along the meridian nearest Paris. They denoted this standard of length with a carefully protected platinum rod. In 1983, scientists refined the SI definition of length with the following statement: "The meter is the length of the path traveled by light in vacuum during a time interval of 1/299,792,458 of a second." Scientists now use the speed of light in vacuum (namely 299,792,458 m/s) as the natural standard for defining the meter.

Other basic units in the SI system are the ampere (A) (a measure of electric current), the candela (cd) (a measure of luminous intensity), the kelvin (K) (a measure of absolute thermodynamic temperature), and the mole (mol) (a measure of the amount of substance). The radian (rad) serves as the basic SI unit of plane angle and the steradian (sr) the basic SI unit of solid angle. There are also numerous supplementary and derived SI units, such as the becquerel (Bq) (a measure of radioactivity) and the newton (N) (a measure of force).

Today, SI measurements play an important role in science and technology throughout the world. However, considerable technical and commercial activity in the United States still involves the use of another set of units, called United States customary system (or American system). This system traces its heritage back to colonial America and the system of weights and measures used in Great Britain in the 18th century.

One major advantage of the SI system over the American system is that SI units employ the number 10 as a base. Therefore, multiples or sub-multiples of SI units are easily reached by multiplying or dividing by 10. The inch (in) and the foot (ft) are the two commonly encountered units of length in the American system. Both originated in medieval England and were based upon human anatomy. The width of a human thumb gave rise to the inch; the human foot, the foot. Through an evolving process of standardization, 12 inches became equal to one foot and three feet equal to one yard (yd).

The SI system uses the kilogram as the basic unit of mass, while the American system uses the pound-mass (lbm). However, the American system also contains another fundamental "pound" unit, the pound-force (lbf), which is actually a unit of force. Both "pounds" were related by an arbitrary decision made centuries ago to create a unit system consisting of four basic units: length, mass, time, and force. To implement this decision, officials declared that "a one pound-mass object exerts (weighs) one pound-force at sea level." Technical novices and professionals alike all too often forget that this arbitrary "pound" equivalency is valid only at sea level on the surface of Earth. The historic arrangement that established the pound-force as a basic (or fundamental) unit rather than a derived unit (based on Newton's second law of motion) persists in the American system and can cause considerable confusion.

In the SI system, a kilogram of mass is a kilogram of mass, whether on Earth or on Mars. Similarly, one pound-mass (lbm) on Earth is still one pound-mass (lbm) on Mars. However, in the American system, if a one pound-mass object "weighs" one pound on Earth, it "weighs" only 0.38 pounds on Mars. What changes the object's weight (as calculated from Newton's second law of motion) is the difference in the local acceleration due to gravity. On Earth, the sea-level acceleration due to gravity (g) is approximately equal to 32.2 feet per second per second (ft/s^2) (about 9.8 m/s^2), while on the surface of Mars, the local acceleration due to gravity (g_{Mars}) is only 12.2 ft/s^2 (3.73 m/s^2).

Scientists use SI units in their work to avoid confusion and ambiguity. Many people find it helpful to remember that the comparable SI unit for the pound-mass is the kilogram (kg), namely 1 kg = 2.205 lbm, while the comparable SI unit for the pound-force is called the newton (N), namely 1 N = 0.2248 lbf.

ORDINARY MATTER VERSUS DARK MATTER

It has taken the collective thinking of some of the very best minds in human history many millennia to recognize the key scientific fact that all matter in the observable universe consists of combinations of different atoms, drawn from a modest collection of about 100 different chemical elements. (See appendix.) The term *observable universe* is intentionally used here, because scientists are now trying to understand and characterize an intriguing invisible form of matter they call dark matter.

This book deals primarily with the characteristics and properties of ordinary (or baryonic) matter. Scientists define *ordinary matter* as matter

HUNTING DARK MATTER
WITH THE *HUBBLE SPACE TELESCOPE*

Scientists currently think dark matter exceeds the total amount of ordinary (or baryonic) matter in the universe by a factor of five or more. At present, the most significant scientific fact known about dark matter is that it exerts a detectable gravitational influence within clusters of galaxies.

Using data from NASA's *Hubble Space Telescope (HST)*, astronomers were able to observe how dark matter behaves during a gigantic collision between two galaxy clusters. The colossal cosmic wreck created a ripple of dark matter. Cleverly, scientists generated a map of the suspected dark matter distribution (shown here) by making very careful observations of how gravitational lensing affected the light from more distant galaxies. Einstein's general theory of relativity predicted gravitational lensing, an intriguing imaging phenomenon caused by the bending, distortion, and magnification of the light from distant cosmic sources as that light passes through less distant gravitational fields.

A *Hubble Space Telescope* (HST) composite image that shows a ghostly, deep blue "ring" of dark matter superimposed upon a visible HST image of galaxy cluster Cl 0024+17. The scientist-generated, deep-blue map of the cluster's dark matter distribution is based on a phenomenon called gravitational lensing. Although "invisible," dark matter is inferred to be present because its gravitational influence bends the incoming light from more distant galaxies. *(NASA, ESA, M.* J. Jee and H. Ford [Johns Hopkins University])

The composite image (shown above) depicts a (scientist-generated) deep blue, ghostly ring of dark matter in the galaxy cluster Cl 0024+17. The ring-like structure of deep blue corresponds to the cluster's dark matter distribution.

(continues)

(continued)

To create this composite image, NASA scientists carefully superimposed the shadowy deep blue–colored map on top of a more familiar, visible image of the same galaxy cluster.

The approach taken by the scientists is somewhat analogous to how fictional authorities used the invisible man's telltale footsteps in the snow to hunt down the rogue scientist in the 1933 motion picture version of H. G. Wells's famous science fiction story. Wells wrote the original novel, entitled *The Invisible Man,* back in 1897.

The discovery of the ring of dark matter in galaxy cluster Cl 0024+17 represents some of the strongest evidence that testifies to the existence of dark matter. Since the 1930s, astronomers and astrophysicists have suspected the presence of some gravitationally influential, invisible substance holding clusters of galaxies together. Scientists estimated that the masses associated with the visible stars in such clusters were simply not enough to generate the gravitational forces needed to keep the whirling galactic clusters together.

composed primarily of baryons. Baryons are typically the heavy nuclear particles formed by the union of three quarks, such as protons and neutrons. (See chapter 9.) Almost everything a person encounters in daily life consists of atoms composed of baryonic matter. Note that beyond this section, whenever the word *matter* appears in the book, it means ordinary, (baryonic) matter unless specifically indicated otherwise.

Scientists define *dark matter* as matter in the universe that cannot be observed directly because it emits very little or no electromagnetic radiation and experiences little or no directly measurable interaction with ordinary matter. They also refer to dark matter as nonbaryonic matter. This means that dark matter does not consist of baryons, as does ordinary matter. So what is dark matter? Today, no one really knows. While not readily observable here on Earth, dark matter nevertheless exerts a very significant large-scale gravitational influence within the universe.

Starting in the 1930s, astronomers began suspecting the presence of dark matter (originally called *missing mass*) because of the observed velocities and motions of individual galaxies in clusters of galaxies.

Astronomers define a *galaxy cluster* (or *cluster of galaxies*) as an accumulation of galaxies (from 10 to 100s or even a few thousand members) that lie within a few million light-years of each other and are bound by gravitation. Without the presence of this postulated dark matter, the galaxies in a particular cluster would have drifted apart and escaped from each other's gravitational grasp long ago. Physicists are now engaged in a variety of research efforts to better define and understand the phenomenon of dark matter.

STATES OF MATTER

One of the more useful ways scientists have devised to classify ordinary matter is to identify the basic physical forms, or states, of matter. They sometimes refer to the states of matter as the phases of matter.

The three most familiar states of matter are solid, liquid, and gas. In general, a solid occupies a specific, fixed volume and retains its shape. A liquid also occupies a specific volume but is free to flow and assume the shape of the portion of the container it occupies. A gas has neither a definite shape nor a specific volume. Rather, it quickly fills the entire volume of a closed container. Unlike solids and liquids, gases can be compressed quite easily. Scientists treat solids and liquids as *condensed matter* and gases as *uncondensed matter.*

When temperatures get sufficiently high, another state of matter occurs. Scientists call this fourth state of matter *plasma*. The glowing neon gas in a colorful sign is one example of plasma encountered here on Earth. The plasma arc cutting torch used in metal welding shops is another. Finally, as temperatures get very low and approach absolute zero, scientists encounter a fifth state of matter called the Bose-Einstein condensate (BEC).

In many scientific and technical applications, scientists and engineers have found it helpful to discuss matter in terms of macroscopic (or bulk) properties. At other times, the microscopic (or molecular) perspective provides more meaningful insight. This book uses both perspectives to discuss the nature and behavior of matter. Scientists often compare physical and chemical properties in their efforts to fully characterize substances. A physical property of a substance is a measurable behavior or physical characteristic. Some of the more common physical properties used by scientists and engineers are mass, density, temperature, structure, and hardness. Other physical properties include a substance's melting point and boiling point. Heat capacity, thermal conductivity, electric conductivity, solubility, and color represent still other physical properties.

The chemical properties of a substance describe how that particular substance reacts with other substances. Viewed on a microscopic scale, the atoms of the substance under study experience change with respect to their interaction with neighboring atoms. Electron trading among neighboring atoms makes or breaks bonds. Changes in ionization state (from electrically neutral to positive or negative) may also take place at the atomic level. As a result of their chemical properties, some substances experience corrosion, others experience combustion (burning) or explosive decomposition, while others (such as helium) remain inert and refuse to interact with other materials in their environment.

Scientists regard a physical change of a substance as one that involves a change in the physical appearance of that particular sample of matter, but the matter that experiences the physical change does not experience any change in its composition or chemical identity. Consider adding heat to an uncovered pot of water on a stove. The water eventually starts to boil and experiences a physical change by becoming a hot gas, called steam. Scientists say that the liquid water experienced a physical change of state when it transformed to steam, but during this process, the steam retained the chemical identity of water (H_2O).

During a chemical change, the original substance changes its chemical identity and experiences a change in composition. Consider another container of water through which a scientist passes a (direct) electric current by means of a battery with platinum (Pt) electrodes. The water molecules in the container now experience a chemical reaction and break down (dissociate) into hydrogen (H_2) and oxygen (O_2) gas molecules. These gas molecules then begin to accumulate at the cathode and anode, respectively. Scientists often add salt to the water to promote the process of water electrolysis. Water electrolysis is the conversion of electrical energy into chemical energy in the form of hydrogen, with oxygen appearing as a useful by-product of the process. The reaction is as follows: $H_2O + \textit{electricity} \rightarrow H_2 + \frac{1}{2} O_2$.

Although scientific characterization of matter is very important, such activities cannot answer a fundamental question that has challenged people since antiquity: Where did matter originally come from? Specifically, what is the origin of the atoms that make up the various chemical elements? To begin to answer this challenging question, scientists had to travel back about 13.7 billion years to the very beginning of time and space. Once there, they had to think about what happened during the fleeting wisps of time immediately after the universe was created in a gigantic ancient explosion called the big bang.

CHAPTER 2

The Origin of Matter

This chapter discusses the origin of ordinary (or baryonic) matter within the context of big bang cosmology. A brief mention is also made of the phenomenon of dark energy and how it, along with dark matter, has helped define the currently observed universe.

THE BIG BANG

The big bang is a widely accepted theory in contemporary cosmology concerning the origin and evolution of the universe. According to the big bang cosmological model, about 13.7 billion years ago, there was an incredibly powerful explosion that started the present universe. Before this ancient explosion, matter, energy, space, and time did not exist. All of these physical phenomena emerged from an unimaginably small, infinitely dense object, which physicists call the *initial singularity*. Immediately after the big bang event, the intensely hot universe began to expand and cool. As the temperature dropped, matter and energy experienced several very interesting transformations. Astrophysical observations indicate that the universe has been expanding ever since, but at varying rates of acceleration due to the continual cosmic tug of war between the mysterious pushing force of dark energy (discussed shortly) and the gravitational pulling force of matter—both ordinary matter and dark matter.

High-energy particle experiments and astrophysical observations lend direct support to the big bang model. The premier observation took place in 1964, when the German-American physicist Arno Allen Penzias

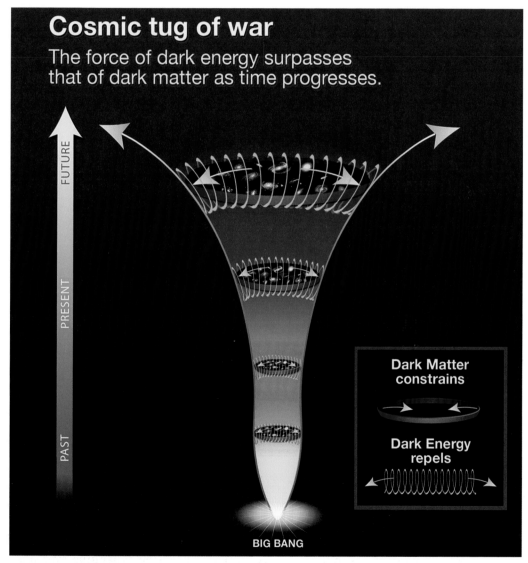

Cosmic tug of war

The force of dark energy surpasses that of dark matter as time progresses.

FUTURE

PRESENT

PAST

Dark Matter constrains

Dark Energy repels

BIG BANG

This diagram suggests that the overall history of the universe can be envisioned as a cosmic tug of war between the gravitational pull of dark matter (the universe's major form of matter) and the push of dark energy (a mysterious phenomenon of nature). This process began at least 9 billion years ago, well before dark energy gained the upper hand and began accelerating the expansion of the universe. *(NASA, ESA, and A. Feild [STScI])*

(1933–) and the American physicist Robert Woodrow Wilson (1936–) detected the cosmic microwave background (CMB). Physicists regard the CMB as the remnant radiation signature of an intensely hot, young universe. Although their discovery occurred quite by accident, their pioneering work provided the first direct evidence that the universe had an explosive beginning.

Prior to the discovery of the CMB, the majority of scientists were philosophically comfortable with a steady state model of the universe—a model postulating that the universe had no beginning and no end. Within this steady state model, scientists assumed matter was being created continually (by some undefined mechanism) to accommodate the universe's observed expansion. The assumption was especially popular, because it allowed the universe to maintain an essentially unchanging appearance over time.

In 1949, British astronomer Sir Fred Hoyle (1915–2001), who helped develop and strongly advocated the steady state model, coined the term *big bang*. Hoyle intended the term to be a derisive expression against fellow

Temperature fluctuations of the cosmic microwave background. The average temperature is 2.725 kelvins (K), and the colors represent the tiny temperature fluctuations. Red regions are warmer, and blue regions are colder by about 0.0002 K. This full sky map of the heavens is based on five years of data collection by NASA's *Wilkinson Microwave Anisotropy Probe* (*WMAP*). *(NASA/WMAP Science Team)*

astronomers who favored big bang cosmology. Unfortunately, Hoyle lost the intellectual battle. His intended derogatory use of the term *big bang* backfired when the expression immediately gained favor among those competing scientists who warmly embraced the term as a snappy way to succinctly explain their new theory of cosmology. Detailed observations of the CMB proved that the universe did, indeed, have an explosive beginning about 13.7 billion years ago. As evidence mounted in favor of big bang cosmology, the vast majority of scientists abandoned the steady state universe model.

Near the end of the 20th century, big bang cosmologists had barely grown comfortable with a gravity-dominated, expanding universe model when another major cosmological surprise popped up. In 1998, two competing teams of astrophysicists independently observed that the universe was not just expanding but expanding at an increasing rate.

Dark energy is the generic name that astrophysicists gave to the unknown cosmic force field thought to be responsible for the acceleration in the rate of expansion of the universe. In 1929, the American astronomer Edwin P. Hubble (1889–1953) proposed the concept of an expanding universe. He suggested that his observations of Doppler-shifted wavelengths of the light from distant galaxies indicated that these galaxies were receding from Earth with speeds proportional to their distance.

In the late 1990s, scientists performed systematic surveys of very distant Type Ia (carbon detonation) supernovas (discussed shortly). They observed that instead of slowing down (as might be anticipated if gravitation were the only significant force at work in cosmological dynamics), the rate of recession (that is, the redshift) of these very distant objects appeared to actually be increasing. It was almost as if some unknown force were neutralizing or canceling the attraction of gravity. Such startling observations proved controversial and very inconsistent with the then-standard gravity-only models of an expanding universe within big bang cosmology. Despite fierce initial resistance within the scientific community, these perplexing observations eventually gained acceptance. Today, carefully analyzed and reviewed Type Ia supernova data indicate that the expansion of the universe is accelerating—a dramatic conclusion that tossed modern cosmology into as much turmoil as Hubble's observations had caused some 70 years earlier.

Physicists do not yet have an acceptable answer as to what phenomenon is causing this accelerated expansion. Some scientists revisited the cosmological constant (symbol Λ). Einstein inserted this concept into his original general relativity theory to make his theory of gravity describe a static universe, that is, a nonexpanding one that has neither a beginning nor an end. After boldly introducing the cosmological constant as representative of some mysterious force associated with empty space capable of balancing or even resisting gravity, Einstein abandoned the idea. Hubble's announcement of an expanding universe provided the intellectual nudge that encouraged his decision. Afterward, Einstein referred to the notion of a cosmological constant as "my greatest failure."

Physicists are now revisiting Einstein's concept and suggesting that there is possibly a vacuum pressure force (a recent name for the cosmological constant) that is inherently related to empty space but that exerts its influence only on a very large scale. The influence of this mysterious force would have been negligible during the very early stages of the universe following the big bang event but would later manifest itself and serve as a major factor in cosmological dynamics. Since such a mysterious force is neither required nor explained by any of the currently known laws of physics, scientists do not yet have a clear physical interpretation of what this gravity-resisting force means.

Dark energy does not appear to be associated with either matter or radiation. In general relativity theory, both energy density (energy per unit volume) and pressure contribute to the strength of gravity. Ordinary (baryonic) matter exerts no pressure of cosmological importance. Dark matter represents a form of exotic, nonbaryonic matter that interacts only weakly with ordinary matter by means of gravity. Although not yet detected directly in particle physics experiments, scientists infer dark matter's existence by its observed large-scale gravitational influence on galactic clusters. Scientists currently think that dark matter does not give rise to any cosmologically significant pressure.

Radiation is composed of massless (or nearly massless) particles, such as photons and neutrinos. Scientists believe that this equivalent form of matter (recall Einstein's mass-energy formula) is characterized as having a large positive pressure. Finally, dark energy is possibly a truly bizarre form of matter, or perhaps a strange property of the vacuum of space itself—a phenomenon arising from the complete absence of ordinary or dark matter. Whatever dark energy really is, it appears to be the only equivalent form of matter characterized by a large negative pressure. Within general relativity, a large negative pressure would act against gravity and cause the rate of expansion of the universe to increase. One of the primary challenges facing scientists in this century is to accurately characterize the cosmic relationships between ordinary matter, dark matter, radiation, and dark energy.

The magnitude of the acceleration in cosmic expansion suggests that the amount of dark energy in the universe actually exceeds the total mass-

MASS–ENERGY

In 1905, while working for the Swiss federal patent office in Bern, Albert Einstein wrote a technical paper entitled "On the Electrodynamics of Moving Bodies." He used this paper to introduce his special theory of relativity—a theory that deals with the laws of physics as seen by observers moving relative to one another at constant velocity. Einstein stated the first postulate of special relativity as "The speed of light (c) has the same value for all (inertial reference frame) observers, independent and regardless of the motion of the light source or the observers." His second postulate of special relativity proclaimed "All physical laws are the same for all observers moving at constant velocity with respect to each other."

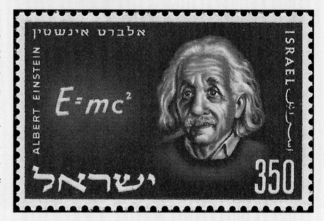

Appearing with Einstein on this 1956 stamp from Israel is his famous mass-energy equivalence equation, arguably the most well known equation in all of physics. *(Author)*

From the special theory of relativity, the theoretical physicist concluded that only a zero rest mass particle, such as a photon, could travel at the speed of light. A major consequence of special relativity is the equivalence of mass and energy. Einstein's famous mass-energy formula, expressed as $E = mc^2$, provides the energy equivalent of matter and vice versa. Among its many important physical insights, this equation was the key that scientists needed to understand energy release in such important nuclear reactions as fission, fusion, radioactive decay, and matter-antimatter annihilation.

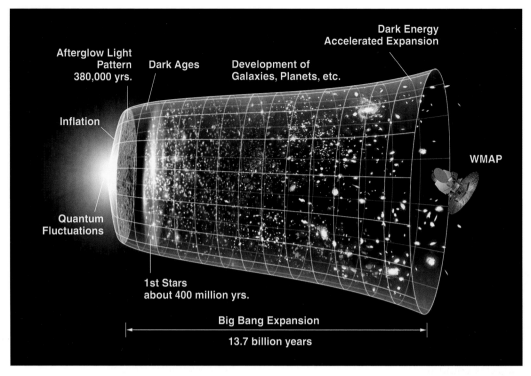

Afterglow Light
Pattern
380,000 yrs.

Dark Ages

Development of
Galaxies, Planets, etc.

Dark Energy
Accelerated Expansion

Inflation

WMAP

Quantum
Fluctuations

1st Stars
about 400 million yrs.

Big Bang Expansion
13.7 billion years

Time line of the universe *(NASA/*WMAP *Science Team)*

energy equivalence of ordinary matter, dark matter, and radiation by a considerable margin. The apparent dominant presence of dark energy implies that the universe will continue to expand forever.

The accompanying illustration depicts the evolution of the universe over the past 13.7 billion years, from the big bang event to the present day (symbolized by NASA's *Wilkinson Microwave Anisotropy Probe [WMAP]* spacecraft). The far left side of this illustration shows the earliest moment scientists can now investigate—the time when an unusual period of *inflation* (discussed shortly) produced a burst of exponential growth in the young universe. (NASA illustrators have portrayed size by the vertical extent of the grid.) For the next several billion years, the expansion of the universe gradually slowed down as the matter in the universe (both ordinary and dark) tugged on itself through gravitational attraction. More recently, the expansion of the universe started to speed up again as the repulsive effects of dark energy became dominant.

The afterglow light seen by NASA's *WMAP* spacecraft was emitted about 380,000 years after the rapid expansion event called inflation and has traversed the universe largely unimpeded since then. When scientists examined the CMB carefully, they discovered telltale signatures imprinted on the afterglow light, revealing information about the conditions of the very early universe. The imprinted information helps scientists understand how primordial events caused later developments (such as the clustering of galaxies) as the universe expanded.

When scientists peer deep into space with their most sensitive instruments, they also look far back in time, eventually viewing the moment when the early universe transitioned from an opaque gas to a transparent gas. Beyond that transition point, any earlier view of the universe remains obscured. The CMB serves as the very distant wall that surrounds and delimits the edges of the observable universe. Scientists can observe cosmic phenomena only back to the remnant glow of this primordial hot gas from the big bang. The afterglow light experienced Doppler shift to longer wavelengths due to the expansion of the universe. The CMB currently resembles emissions from a cool dense gas at a temperature of only 2.725 K.

The very subtle variations observed in the CMB are challenging big bang cosmologists. They must explain how the clumpy structures of galaxies could have evolved from a previously assumed smooth (that is, uniform and homogeneous) big bang event. Launched on June 30, 2001, NASA's *WMAP* spacecraft made the first detailed full sky map of the CMB, including high-resolution data of those subtle CMB fluctuations that represent the primordial seeds that germinated into the cosmic structure scientists observe today. The patterns detected by *WMAP* represent tiny temperature differences within the very evenly dispersed microwave light bathing the universe—a light that now averages a frigid 2.725 K. The subtle CMB fluctuations have encouraged physicists to make modifications in the original big bang hypothesis. One of these modifications involves a concept they call *inflation*. During inflation, vacuum state fluctuations gave rise to the very rapid (exponential), nonuniform expansion of the early universe.

In 1980, American physicist Alan Harvey Guth (1947–   ) proposed inflation in order to solve some of the lingering problems in cosmology that were not adequately treated in standard big bang cosmology. As Guth and, later, other scientists suggested, between 10^{-35} and 10^{-33} s after the big bang, the universe actually expanded at an incredible rate—far in excess of the speed of light. In this very brief period, the universe increased in size by at least a factor of 10^{30}, growing from an infinitesimally small subnuclear-sized dot to a region about 9.84 feet (3 m) across. By analogy, imagine a grain of very fine sand becoming the size of the presently observable universe in one-billionth (10^{-9}) the time it takes light to cross the nucleus of an atom. During inflation, space itself expanded so rapidly that the distances between points in space increased greater than the speed of light. Scientists suggest that the slight irregularities they now observe in the CMB are evidence (the fossil remnants, or faint ghosts) of the quantum fluctuations that occurred as the early universe inflated.

Astrophysicists use three measurable signatures that strongly support the notion that the present universe evolved from a dense, nearly featureless hot gas, just as big bang theory suggests. These scientifically measurable signatures are the expansion of the universe, the abundance of the primordial light elements (hydrogen, helium, and lithium), and the CMB radiation. Hubble's observation in 1929 that galaxies were generally receding (when viewed from Earth) provided scientists their first tangible clue that the big bang theory might be correct. The big bang model also suggests that hydrogen, helium, and lithium nuclei should have fused from the very energetic collisions of protons and neutrons in the first few minutes after the big bang. (Scientists call this process *big bang nucleosynthesis.*) The observable universe has an incredible abundance of hydrogen and helium.

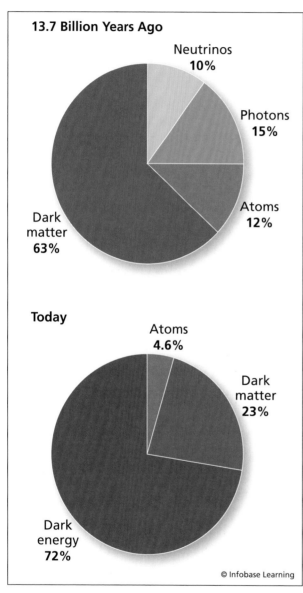

13.7 Billion Years Ago

Neutrinos 10%

Photons 15%

Atoms 12%

Dark matter 63%

Today

Atoms 4.6%

Dark matter 23%

Dark energy 72%

© Infobase Learning

Content of the past and present universe based upon evaluation of five years of data from NASA's *WMAP* spacecraft. (Note that *WMAP* data are accurate only to two digits, so the total appearing for today's universe is not exactly 100 percent.) *(NASA/WMAP Science Team)*

Finally, within the big bang model, the early universe should have been very, very hot. Scientists regard the CMB (first detected in 1964) as the remnant heat left over from an ancient fireball.

The importance of the CMB to modern cosmology cannot be understated. On March 7, 2008, NASA released the results of a five-year investigation of the oldest light in the universe. Based on a careful evaluation of *WMAP* data, scientists were able to gain incredible insight into the past and present contents of the universe. The *WMAP* data (see the accompanying illustration) revealed that the current contents of the universe include 4.6 percent atoms—the building blocks of stars, planets, and people. In contrast, dark matter constitutes 23 percent of the universe. Finally, 72 percent of the current universe is composed of dark energy, which acts as a type of matter-repulsive, antigravity phenomenon.

Scientists speculate that dark energy is distinct from dark matter and is responsible for the present-day acceleration of universal expansion. The NASA-developed illustration of the universe's contents reflects the current limits inherent in the *WMAP* spacecraft's ability to estimate the amount of dark matter and dark energy by observing the CMB. Note that *WMAP* data are accurate to two digits, so the total appearing in the contents of today's universe is not exactly 100 percent. Despite these minor limitations, the results are rather startling. The contents of the current universe and the early universe (about 380,000 years after the big bang) are quite different from each other. This suggests the persistent influence of the cosmic tug of war between energy (radiant and dark) and matter (baryonic and nonbaryonic). Subsequent portions of this book deal with how human beings searched for the meaning of substance and eventually came to understand ordinary (baryonic) matter.

HOW HYDROGEN AND HELIUM FORMED IN THE ANCIENT INFERNO

Contemporary scientific measurements estimate that hydrogen makes up approximately 73 percent of the mass of ordinary matter in the observable universe. Helium makes up about 25 percent of the mass, and everything else constitutes only 2 percent of the mass. While the relative abundances of the chemical elements beyond hydrogen and helium appear quite low in these estimates, it is important to realize that most of the atoms here on Earth, including those in a person's body, are part of this small portion of ordinary matter. The primordial low-mass nuclei (hydrogen, deuterium, helium-3, helium-4, and lithium-7 [only trace quantities]) were forged in the crucible of the very early universe just minutes after the big bang event. At that time, temperatures were about 10^9 K, and big bang nucleosynthesis took place. These low-mass nuclei owe their existence and elemental heritage to the intensely hot and very dense conditions that accompanied the birth of the universe.

Although some of the details remain a bit sketchy (due to limitations in today's physics), this section explains where the elements hydrogen and helium came from. The next section describes how all the other chemical elements found here on Earth (and elsewhere in the universe) were created by nuclear reactions during the life and death of ancient stars.

Physicists call the incredibly tiny time interval between the big bang event and 10^{-43} s the *Planck time* or the *Planck era* in honor of Max Planck, who proposed quantum theory. Current levels of science are inadequate to describe the extreme conditions of the very young universe during the Planck era. Scientists can only speculate that during the Planck era the very young universe was extremely small and extremely hot—perhaps at a temperature greater than 10^{32} K. At such an unimaginably high temperature, the four fundamental forces of nature were most likely merged into one unified super force. Limitations imposed by the uncertainty principle of quantum mechanics prevent scientists from developing a model of the universe between the big bang event and the end of Planck time. This shortcoming forged an interesting intellectual alliance between astrophysicists (who study the behavior of matter and energy in the large-scale universe) and high-energy nuclear particle physicists (who investigate the behavior of matter and energy in the subatomic world).

In an effort to understand conditions of the very early universe, researchers are trying to link the physics of the very small (as described by quantum mechanics) with the physics of the very large (as described by Einstein's general relativity theory). What scientists hope to develop is a new realm of physics that they call *quantum gravity*. If developed, quantum gravity would be capable of treating Planck era phenomena. An important line of contemporary investigation involves the use of very powerful particle accelerators here on Earth to briefly replicate the intensely energetic conditions found in the early universe. (See chapters 8 and 9.) Particle physicists share the results of their very energetic particle-smashing experiments with astrophysicists and cosmologists, who are looking far out to the edges of the observable universe for telltale signatures that suggest how the universe evolved to its present-day state from an initially intense inferno.

Particle physicists currently postulate that the early universe experienced a specific sequence of phenomena and phases following the big bang explosion. They generally refer to this sequence as the *standard cosmological model*. Right after the big bang, the present universe was at an incredibly high temperature. Physicists suggest the temperature was at the unimaginable value of about 10^{32} K. During this period, which they sometimes call the *quantum gravity epoch*, the force of gravity, the strong force, and a composite electroweak force all behaved as a single unified force. Scientists further postulate that, at this time, the physics of elementary particles and the physics of space-time were one and the same. In an effort to adequately describe quantum gravity, many physicists are now searching for a theory of everything (TOE).

FUNDAMENTAL FORCES IN NATURE

At present, physicists recognize the existence of four fundamental forces in nature: gravity, electromagnetism, the strong force, and the weak force. Gravity and electromagnetism are part of a person's daily experiences. These two forces have an infinite range, which means they exert their influence over great distances. The two stars in a binary star system, for example, experience mutual gravitational attraction even though they may be a light-year or more distant from each other.

The other two forces, the strong force and the weak force, operate within the realm of the atomic nucleus and involve elementary particles. These forces lie beyond a person's normal experiences and remained essentially unknown to the physicists of the 19th century despite their good understanding of classical physics—including the universal law of gravitation and the fundamental principles of electromagnetism. The strong force operates at a range of about 10^{-15} m and holds the atomic nucleus together. The weak force has a range of about 10^{-17} m and is responsible for processes such as beta decay that tear nuclei and elementary particles apart. What is important to recognize here is that whenever anything happens in the universe—that is, whenever an object experiences a change in motion—the event takes place because one or more of these fundamental forces is involved.

At Planck time (about 10^{-43} s after the big bang), the force of gravity assumed its own identity. With a temperature estimated to be about 10^{32} K, the entire spatial extent of the universe at that moment was less than the size of a proton.

Scientists call the ensuing period the *grand unified epoch.* While the force of gravity functioned independently during this period, the strong force and the composite electroweak force remained together and continued to act as a single force. Today, physicists apply various grand unified theories (GUTs) in their efforts to model and explain how the strong force and the electroweak force functioned as one during this particular period in the early universe.

About 10^{-35} s after the big bang, the strong force separated from the electroweak force. By this time, the expanding universe had cooled to about 10^{28} K. Physicists call the period between about 10^{-35} s and 10^{-10} s the *electroweak force epoch.* During this epoch, the weak force and the electromagnetic force became separate entities, as the composite electroweak force disappeared. From that time forward, the universe contained four fundamental natural forces, namely, the force of gravity, the strong force, the weak force, and the electromagnetic force.

Following the big bang and lasting up to about 10^{-35} s, there was no distinction between quarks and leptons. All the minute particles of matter were similar. However, during the electroweak epoch, quarks and leptons became distinguishable. This transition allowed quarks and antiquarks to eventually become hadrons, such as neutrons and protons, as well as their antiparticles. At 10^{-4} s after the big bang, in a period scientists call the *radiation-dominated era,* the temperature of the universe cooled to 10^{12} K. By that time, most of the hadrons had disappeared due to matter–antimatter annihilations. The surviving protons and neutrons represented only a small fraction of the total number of particles in the early universe, the majority of which were leptons, such as electrons, positrons, and neutrinos, but, like most of the hadrons before them, the majority of the leptons also quickly disappeared because of matter–antimatter interactions.

At the beginning of the period scientists call the *matter-dominated era,* which occurred some three minutes (or 180 s) following the big bang, the expanding universe had cooled to a temperature of 10^9 K, and low-mass nuclei, such as deuterium, helium-3, helium-4, and (a trace quantity of) lithium-7 began to form due to very energetic collisions in which the reacting particles joined or fused. Once the temperature of the expanding universe dropped below 10^9 K, environmental conditions no longer favored the fusion of low-mass nuclei, and the process of big bang nucleosynthesis ceased. Physicists point out that the observed amount of helium in the universe agrees with the theoretical predictions concerning big bang nucleosynthesis.

Finally, when the expanding universe reached an age of about 380,000 years, the temperature had dropped to about 3,000 K, allowing electrons and nuclei to combine and form hydrogen and helium atoms. Within the period of atom formation, the ancient fireball became transparent and continued to cool from a hot 3,000 K down to a frigid 2.725 K—the currently measured temperature of the CMB.

PEOPLE ARE MADE OF STARDUST

The American astronomer Carl Sagan (1934–96) was fond of stating that "We are made of stardust." The interesting comment means that, except for hydrogen (which formed within minutes after the big bang), all the elements found in the human body came from processes involving the life and death of ancient stars. This section briefly describes the cosmic heritage of the chemical elements.

There were not any stars shining when the universe became transparent some 380,000 years after the big bang, so scientists refer to the ensuing period as the cosmic *dark ages*. During the cosmic dark ages, very subtle variations (inhomogeneities) in density allowed the attractive force of gravity to form great clouds of hydrogen and helium. Gravity continued to subdivide these gigantic gas clouds into smaller clumps, enabling the formation of the first stars. As many brilliant new stars appeared, they gathered into galaxies. Astronomers currently estimate that about 400,000 years after the big bang, the cosmic environmental conditions were right for a rapid rate of star formation. Soon after, the cosmic dark ages became illuminated by the radiant emissions of millions of young stars.

Gravitational attraction slowly gathered clumps of hydrogen and helium gas into new stars. The very high temperatures in the cores of massive early stars supported the manufacture of heavier nuclei up to and including iron by means of a process scientists call *nucleosynthesis*. Elements heavier than iron were formed in a bit more spectacular fashion. Neutron-capture processes deep inside highly evolved massive stars and subsequent supernova explosions at the ends of their relatively short lifetimes synthesized all the elements beyond iron.

A star forms when a giant cloud of mostly hydrogen gas, perhaps light-years across, begins to contract under its own gravity. Over millions of years, this clump of hydrogen and helium gas eventually collects into a giant ball that is hundreds of thousands of times more massive than Earth. As the giant gas ball continues to contract under its own gravitational influence, an enormous pressure arises in its interior. The

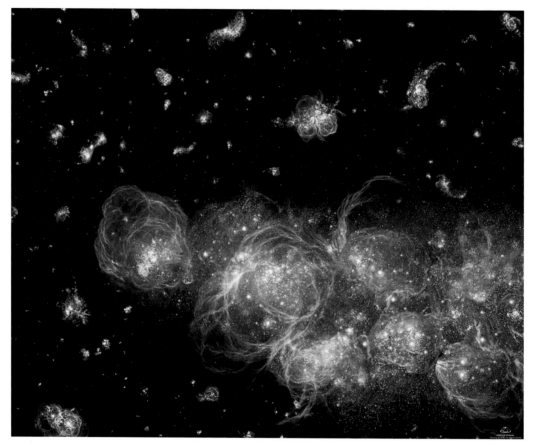

This is an artist's rendering of how the early universe (less than 1 billion years after the big bang) must have looked when it went through a very rapid onset of star formation, converting primordial hydrogen into a myriad of stars at an unprecedented rate. Regions filled with newborn stars began to glow intensely. Brilliant new stars began to illuminate the "dark ages." The most massive of these early stars self-detonated as supernovas, creating and spreading chemical elements throughout the fledgling universe. Analysis of *Hubble Space Telescope* data supports the hypothesis that the universe's first stars appeared in an eruption of star formation rather than at a gradual pace. *(NASA, K. Lanzetta [SUNY], and Adolf Schaller [STScI])*

increase in pressure at the center of the protostar is accompanied by an increase in temperature. When the center of the contracting gas ball reaches a minimum temperature of about 10 million K, the hydrogen nuclei acquire sufficient velocities to experience collisions that support nuclear fusion reactions. (The central temperature of the Sun's core is approximately 15 million K.) This is the moment a new star is born. The process of nuclear fusion releases a great quantity of energy at the center of the young star. Once thermonuclear burning begins in the star's core, the energy released counteracts the continued contraction of stellar mass by gravity. As the inward pull of gravity exactly balances the outward radiant pressure from thermonuclear fusion reactions in the core, the contracting ball of gas becomes stable. Ultimately, the energy released in fusion flows upward to the star's outer surface, and the new star radiates its energy into space.

Size definitely matters in stellar astronomy. Stars come in a variety of sizes, ranging from about one-10th to 60 (or more) times the mass of the Sun. It was not until the mid-1930s that scientists began to recognize that the process of nuclear fusion takes place in the interiors of all normal stars and fuels their enormous radiant energy outputs. Scientists use the term *nucleosynthesis* to describe the complex process of how different size stars create different elements through nuclear fusion reactions and various alpha particle (helium nucleus) capture reactions. All stars on the main sequence use thermonuclear reactions to convert hydrogen into helium, liberating energy in the process. The initial mass of a star determines not only how long it lives (as a main sequence star), but also how it dies.

Small- to average-mass stars share the same fate at the end of their relatively long lives. At birth, low-mass stars begin their stellar lives by fusing hydrogen into helium in their cores. This process generally continues for billions of years until there is no longer enough hydrogen in a particular stellar core to fuse into helium. Once hydrogen burning stops, so does the release of the thermonuclear energy that produced the outward radiant pressure, which counteracted the relentless inward attraction of gravity.

At this point in its life, the small star begins to collapse inward. Gravitational contraction causes an increase in temperature and pressure. Any hydrogen remaining in the star's middle layers soon becomes hot enough to undergo thermonuclear fusion into helium in a shell around the dying star's core. The release of fusion energy in this shell enlarges the star's outer layers, causing the star to expand far beyond its previous dimensions. The

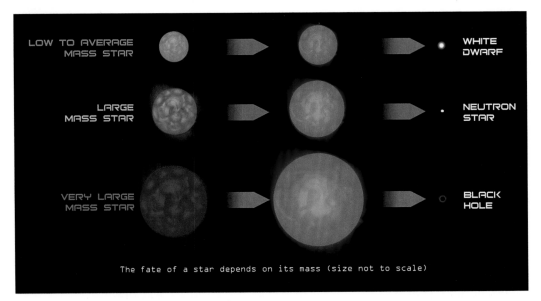

LOW TO AVERAGE MASS STAR → → WHITE DWARF

LARGE MASS STAR → → NEUTRON STAR

VERY LARGE MASS STAR → → BLACK HOLE

The fate of a star depends on its mass (size not to scale)

A star's mass determines its ultimate fate. (Not drawn to scale) *(NASA/CXC)*

expansion process cools the outer layers of the star, transforming them from brilliant white hot or bright yellow in color to a shade of dull glowing red. Astronomers call a low-mass star at this point in its life cycle a *red giant.*

Gravitational attraction continues to make the low-mass star collapse, until the pressure in its core reaches a temperature of about 100 $\times 10^6$ K. This very high temperature is sufficient to allow the thermonuclear fusion of helium into carbon. The fusion of helium into carbon now releases enough energy to prevent further gravitational collapse—at least until the helium runs out. This stepwise process continues until oxygen is fused. When there is no more material to fuse at the progressively increasing high-temperature conditions within the collapsing core, gravity again exerts it relentless attractive influence on matter. This time, the heat released during gravitational collapse causes the outer layers of the low-mass star to blow off, creating an expanding symmetrical cloud of material that astronomers call a *planetary nebula.* The planetary nebula may contain up to 10 percent of the dying star's mass. The explosive blow off process is very important, because it disperses into space some of the low-mass elements (such as carbon) that were manufactured by the small dying star.

The final collapse that causes the small star to eject a planetary nebula also releases energy, but this energy release is insufficient to fuse other elements. The remaining core material continues to collapse until all the atoms are crushed together and only the repulsive force between the electrons counteracts gravity's pull. Physicists refer to this type of very condensed matter as *degenerate matter*. The final compact object is a degenerate matter star, called a *white dwarf*. The white dwarf represents the final phase in the evolution of most low-mass stars, including the Sun.

In some very rare cases, a white dwarf can undergo a gigantic explosion that astrophysicists call a *Type Ia supernova*. This happens when the white dwarf is part of a binary star system and pulls too much matter from its stellar companion. The moment the compact white dwarf can no longer support the addition of incoming mass from its stellar companion, it experiences a new wave of gravitational collapse followed by runaway thermonuclear burning and completely explodes. Nothing is left behind. All the chemical elements created during the lifetime of the low-mass star are now scattered into space in a spectacular detonation.

Astrophysicists define *large stars* as those with more than three to five times the mass of the Sun. Large stars begin their lives in much the same way as small stars—by fusing hydrogen into helium. Because of their size, large stars burn their thermonuclear fuel faster and hotter, typically fusing all the hydrogen in their cores into helium in less than 1 billion years. Once the hydrogen in a large star's core is fused into helium, it becomes a red supergiant, a stellar object similar to a red giant, only much larger. The red supergiant star experiences very high core temperatures during gravitational contraction and begins to fuse helium into carbon, carbon and helium into oxygen, and even two carbon nuclei into magnesium. Thus, through a combination of intricate nucleosynthesis reactions, the red supergiant forms progressively heavier elements, up to and including the element iron. Astrophysicists suggest that the red supergiant forms an onionlike structure, with different elements being fused at different temperatures in layers around the core.

When nuclear reactions begin to fill the core of a red supergiant with iron, the overall thermonuclear energy release in the interior begins to decrease. (Iron nuclei do not undergo energy-liberating nuclear fusion reactions.) Because of this decline, the dying massive star no longer has the internal radiant pressure to resist the attractive force of gravity. A sudden gravitational collapse causes the core temperature to rise to more than 100×10^9 K, smashing the electrons and protons in each iron atom together to form neutrons. The force of gravity now draws this massive collection of neutrons incredibly close together. For about a second, the

This mosaic image is one of the largest ever taken by NASA's *Hubble Space Telescope* of the Crab Nebula in the constellation of Taurus. The image shows a six-light-year-wide expanding remnant of a star's supernova explosion. The orange filaments are the tattered remains of the star and consist mostly of hydrogen. A rapidly spinning neutron star (the crushed, ultradense core of the exploded star) is embedded in the center of the nebula and serves as the dynamo powering the nebula's eerie interior bluish glow. Chinese astronomers witnessed this violent event in 1054 C.E. *(NASA, ESA, J. Hester [Arizona State University])*

neutrons fall very fast toward the highly compressed center of the star. Then, while they are crunching each other, the neutrons abruptly stop. This sudden stop causes the neutrons to recoil violently, and an explosive shockwave travels outward from the highly compressed core. As the shockwave travels from the core, it heats and accelerates the outer layers of material of the red supergiant, causing the majority of the large star's mass to be blown off into space. Astrophysicists call this enormous explosion a *Type II supernova*. The stellar core end product of a Type II supernova is either a neutron star or a black hole.

Astrophysicists suggest that the slow neutron capture process (or *s-process*) produces heavy nuclei up to and including bismuth-209, the most massive naturally occurring nonradioactive nucleus. The flood of neutrons that accompanies a Type II supernova explosion is responsible for the rapid neutron capture process (or *r-process*). It is these rapid neutron capture reactions that form the radioactive nuclei of the heaviest elements found in nature, such as thorium-232 and uranium-238. The violently explosive force of the supernova also hurls stellar-minted elements throughout interstellar space in a dramatic shower of stardust. The expelled stardust eventually combines with interstellar gas. The elementally enriched interstellar gas then becomes available to create a new generation of stars, including those with a family of planets.

About 4.6 billion years ago, our solar system (including planet Earth) formed from just such an elementally enriched primordial cloud of hydrogen and helium gas. All things animate and inanimate here on Earth are the natural by-products of the life and death of massive ancient stars.

CHAPTER 3

The Search
for Substance

The history of civilization is the story of the human mind understanding matter. This chapter presents some of the early intellectual achievements and ancient technical developments that led to the scientific interpretation of substance. Throughout most of human history, the process of understanding the nature of matter was a very gradual one.

MATERIALS AND THE RISE OF MAN

The Paleolithic period (Old Stone Age) represents the longest phase in human development. Archaeologists and anthropologists often divide this large expanse of time into three smaller periods: the Lower Paleolithic period, the Middle Paleolithic period, and the Upper Paleolithic period. Taken as a whole, the most important development that occurred during the Paleolithic period was the evolution of the human species from a near-human apelike creature into modern human beings (called *Homo sapiens*). The process was exceedingly slow, starting about 2 million years ago and ending about 10,000 years ago with the start of the Mesolithic period (Middle Stone Age), an event that coincided with the end of the last ice age.

Any detailed discussion of human evolution stimulates a great deal of controversy, both inside and outside the scientific community. The purpose of this chapter is simply to describe how scientifically credible evidence (such as properly dated bones, artifacts, stone tools, and cave paintings) tells the story of the gradual transition of human beings from nomadic hunter-gathers to farmers and city builders. The latter group established the social units that grew into the first civilizations.

In the ancient Middle East, the earliest civilizations arose along the Nile River (in Egypt) and in Mesopotamia, a Greek word meaning "between rivers." The rivers referred to are the Tigris and Euphrates Rivers, now found in modern Iraq. Historians call Mesopotamia the *Fertile Crescent* or the *Cradle of Civilization*. Other early civilizations appeared in the Indus River Valley of South Asia and in the Yellow River Valley of China. Despite numerous external influences and social transformations, these four ancient civilizations have served as the basis of continuous cultural development in the same geographic locations for all of recorded history. Two other ancient societies, the Minoans on the island of Crete and the native inhabitants of Meso-America (that is, Mexico and Central America), developed civilizations with cities and complex social structures, but their cultures were eventually submerged by invasions and natural disasters.

Scientists suggest that the first civilizations had the following common features: The inhabitants practiced sophisticated levels of agriculture, including the domestication of animals; their people learned to use metals and to make advanced types of pottery and bricks; they constructed cities; they developed complex social structures, including class systems; and they developed various forms of writing. While humankind existed long before these early civilizations appeared, the ability to write made recorded history possible. Anthropologists regard the use of a written language as a convenient division between prehistoric and historic societies.

The periods cited in this section are only approximate and vary significantly from geographic region to geographic region. The Lower Paleolithic period extends from approximately 2 million years ago until about 100,000 B.C.E. and represents the earliest and longest specific period of human development. During this period, early hunter-gatherers learned to use simple stone tools for cutting and chopping. Crude handheld stone axes represented a major advancement. In the Middle Paleolithic period (from about 100,000 B.C.E. to about 40,000 B.C.E.), Neanderthal man lived in caves, learned to control fire, and employed improved stone tools for hunting, including spears with well-sharpened stone points. These nomadic people also learned how to use bone needles to sew furs and

This artist's rendering shows Earth (Europe) about 60,000 years ago. The members of a Neanderthal hunter-gatherer band prepare a meal while huddling around a fire for warmth and security. In the distance, two elder members of the clan gaze intently at the night sky, perhaps puzzled by the wandering, bright red light (the planet Mars). Wooly mammoths and other Pleistocene Age animals appear in the background. *(NASA; artist Randii Oliver)*

animal skins into body coverings. Archaeological evidence suggests these early peoples began to employ fire as a powerful tool not only for warmth and illumination but also to repel dangerous animals, to herd and hunt desirable game, and even to set controlled burns that encouraged the growth of game-attracting, grassy vegetation.

During the Upper Paleolithic period (from about 40,000 B.C.E. to approximately 10,000 B.C.E.), Cro-Magnon man arrived on the scene and displaced Neanderthal man. Cro-Magnon tribal clans initiated more organized and efficient hunting and fishing activities with improved tools and weapons, including carefully sharpened obsidian and flint blades. A great variety of finely worked stone tools, better sewn clothing, the first human constructed shelters, and jewelry (such as bone and ivory necklaces) also appeared. The shaman of Cro-Magnon clans painted colorful animal pictures on the walls of caves. The ancient cave paintings that survive provide scientists some insight into the rituals of Cro-Magnon hunting cultures. One example is the interesting collection of cave paintings that were made sometime between 18,000 B.C.E. and 15,000 B.C.E. in Vallon-Pont-d'Arc, France.

The first important step leading to the understanding of matter was the discovery and use of fire in prehistoric times. A companion intellectual breakthrough by early humans involved the development and use of very simple tools, such as a stick, a sharp rock, and similar objects. These developments were extremely gradual; when they did occur, they typically led to a division of labor within ancient hunting and gathering societies. The survival of a prehistoric clan in the late Upper Paleolithic period not only depended upon the skill of its hunters and gatherers, but also upon the expertise of its toolmakers and fire-keepers.

The Mesolithic period (also called the Middle Stone Age) featured the appearance of better cutting tools, carefully crafted stone points for spears and arrows, and the bow. Depending on the specific geological region, the Mesolithic period began about 12,000 years ago. This era of human development continued until it was replaced by another important technical and social transformation anthropologists call the *Neolithic Revolution*. Scientists believe that humans domesticated the dog during the Mesolithic period and then began using the animal as a hunting companion. The prehistoric human-canine bond continues to this very day.

The Neolithic Revolution (New Stone Age) was the incredibly important transition from hunting and gathering to agriculture. As the last ice age ended about 12,000 years ago, various prehistoric societies in the Middle East (the Nile Valley and Fertile Crescent) and in parts of Asia independently began to adopt crop cultivation. Since farmers tend to stay in one place, the early peoples involved in the Neolithic Revolution began to establish semipermanent and then permanent settlements. Historians often identify this period as the beginning of human civilization. The Latin word *civis* means "citizen" or "a person who inhabitants a city."

The Neolithic Revolution took place in the ancient Middle East starting about 10,000 B.C.E. and continued until approximately 3500 B.C.E. During this period, increasing numbers of people shifted their dependence for subsistence from hunting and gathering to crop cultivation and animal domestication. The growing surplus of food and the control of fire set the stage for the emergence of artisan industries, such as pottery making and metal processing. The transport and marketing of surplus food and artisan goods gave rise to merchants and far-reaching economic activities.

Archaeologists suggest that human beings used the barter system at least 100,000 years ago. As trading activities increased within and between the early civilizations that emerged in the ancient Middle East and around the Mediterranean Sea region, the need for a more efficient means of transferring goods and storing wealth also arose. In about 3000 B.C.E., for example, the people of Mesopotamia began using a unit of weight they called the *shekel* to identify a specific mass of barley (estimated to be about 0.022 lbm [0.010 kg]). The shekel became a generally accepted weight for trading other commodities, such as copper, bronze, and silver. In about 650 B.C.E., the Lydians (an Iron Age people whose kingdom was located in the western portion of modern Turkey) invented money by introducing officially stamped pieces of gold and silver that contained specific quantities of these precious metals. Money has been the lubricant of global economic activity ever since.

As part of the Neolithic Revolution, humans developed the advanced tools and material manipulating skills needed to create and maintain settlements occupied by a few hundred people or more. The city of Jericho (a town near the Jordan River in the West Bank of the Palestinian territories) is an example. With a series of successive settlements ranging from the present back to about 9000 B.C.E., archaeologists regard Jericho as the oldest continuously inhabited settlement in the world.

Civilizations began to form and flourish when farmers produced enough food to support the nonagricultural specialists, who became artisans, merchants, soldiers, and politicians. The artisans, for example, focused on developing the material-processing skills needed for metalworking and for advanced pottery making, further accelerating the growth of civilization.

During the Neolithic (New Stone Age) period, people designed better tools and invented a variety of simple machines. As populations swelled, new tools and simple machines allowed the people of early societies to incorporate different materials in their efforts to modify the surroundings to better suit survival and growth. The natural environment changed as simple dams, irrigation canals, roads, and walled villages appeared. The rise of prosperous ancient civilizations in Mesopotamia is directly attributed to skilled stewardship of water resources, especially the introduction of innovative irrigation techniques.

With the transition to agriculture came food surpluses, the domestication of animals, the production of clothing, the growing use of metals, the expansion of trade, the further specialization of labor, and a variety of interesting social impacts. The rise of civilization spawned the first governments (based on the need to organize human labor and focus wealth in the development of various public projects), the collection of taxes (to pay for government), the first organized military establishments (to protect people and their possessions), and the first schools (to pass on knowledge and technical skills to future generations in a more or less organized fashion). When bountiful economic circumstances permitted, a few people were even able to earn a living by simply thinking or teaching. These individuals became the philosophers, mathematicians, and early scientists, whose accumulated knowledge and new ideas influenced the overall direction and rate of human development. The philosophers and mathematicians of ancient Greece exerted a tremendous influence on the overall trajectory of Western civilization.

The Bronze Age occurred in the ancient Middle East starting about 3500 B.C.E. and lasting until about 1200 B.C.E. During this period, metalworking artisans began to use copper and bronze (a copper and tin alloy) to make weapons and tools. As the demand for copper and bronze products grew, inhabitants of the early civilizations (such as the Minoans on Crete) expanded their search for additional sources of copper and tin. Bronze Age mercantile activities encouraged trade all around the Mediterranean Sea.

About 1200 B.C.E., some civilizations in the ancient Middle East entered the Iron Age, when their metalworking artisans discovered various iron smelting techniques. Civilizations such as the Hittite Empire of Asia Minor enjoyed a

This 1975 Swiss postage stamp depicts three elaborate Bronze Age daggers (produced ca. 1500 to 1800 B.C.E.) that are now at the Lausanne Museum of Archaeology. *(Author)*

METAL OF CYPRUS

In the Bronze Age, the Minoans sponsored settlements on Cyprus and developed copper mines there. Cyprus is the third largest island in the Mediterranean Sea. During the Roman Empire, the Latin phrase *aes Cyprium* meant "metal of Cyprus." The phrase was eventually shortened to *cuprum*—the Latin name for copper and also the source of the chemical element symbol, Cu.

military advantage as their artisans began to produce iron weapons that were stronger than bronze weapons. Later during this period, the very best weapons and tools were those made of steel, an alloy of iron and varying amounts of carbon. Over the next six centuries, the use of metal tools and weapons spread rapidly to Greece, Rome, and other parts of Europe.

By about 1000 B.C.E., human beings in various early civilizations had discovered and were using the following chemical elements (listed here in alphabetical order): carbon (C), copper (Cu), gold (Au), iron (Fe), lead (Pb), mercury (Hg), silver (Ag), sulfur (S), tin (Sn), and zinc (Zn). The use of metals and other materials supported the rise of advanced civilizations all around the Mediterranean Sea, with the Roman Empire serving as an enormous political stimulus and evolutionary endpoint. When the Roman Empire collapsed, western Europe fragmented into many smaller political entities. Progress in materials science continued, but at a much more fractionated and sporadic pace, until the 17th century and the start of the Scientific Revolution.

EARLY VIEWS OF MATTER: EARTH, AIR, WATER, AND FIRE

The Greek philosopher Thales of Miletus (ca. 624–545 B.C.E.) was the first European thinker to suggest a theory of matter. About 600 B.C.E., he postulated that all substances came from water and would eventually turn back into water. Thales may have been influenced in his thinking by the fact that water assumes all three commonly observed states of matter: solid (ice), liquid (water), and gas (steam or water vapor). The evaporation of water under the influence of fire or sunlight could have given him the notion of recycling matter.

Anaximenes (ca. 585–25 B.C.E.) was a later member of the school of philosophy at Miletus. He followed in the tradition of Thales by proposing one primal substance as the source of all other matter. For Anaximenes, that fundamental substance was air. In surviving portions of his writings, he suggests, "Just as our soul being air holds us together, so do breath and air encompass the whole world." He proposed that cold and heat, moisture, and motion made air visible. He further stated that when air was rarified it became diluted and turned into fire. In contrast, the winds, for Anaximenes, represented condensed air. If the condensation process continued, the result was water, with further condensation resulting in the primal substance (air) becoming earth and then stones.

The Greek pre-Socratic philosopher Empedocles (ca. 495–35 B.C.E.) lived in Acragas, a Greek colony in Sicily. In about 450 B.C.E. he wrote the

A diagram describing the interaction of the ancient Greek elements, earth, air, water, and fire, superimposed upon a photograph of a prescribed burn of marsh grasses in a wildlife refuge in California *(U.S. Fish and Wildlife Service; modified by author)*

poem *On Nature*. In this lengthy work (of which only fragments survive), he introduced his theory of a universe in which all matter is made up of four classic elements: earth, air, water, and fire—that periodically combine and separate under the influence of two opposing forces (love and strife). According to Empedocles, fire and earth combined to produce dry conditions, earth blends with water to form cold, water combines with air to produce wet, and air and fire combine to form hot.

In about 430 B.C.E., the Greek natural philosopher Democritus elaborated upon the atomic theory of matter and emphasized that all things consist of changeless, indivisible, tiny pieces of matter called atoms. According to Democritus, different materials consisted of different atoms, which interacted in certain ways to produce the particular properties of a specific material. What is remarkable about ancient Greek atomism is that the idea tried to explain the great diversity of matter found in nature with just a few basic concepts tucked into a relatively simple theoretical framework. Unfortunately, the atomistic theory of matter fell from favor when the more influential Greek philosopher Aristotle rejected the concept.

Starting in about 340 B.C.E., Aristotle embraced and embellished the theory of matter originally proposed by Empedocles. Within Aristotelian cosmology, planet Earth was the center of the universe. Everything within Earth's sphere was composed of a combination of the four basic elements: earth, air, water, and fire. Aristotle suggested that objects made of these basic four elements are subject to change and move in straight lines. However, heavenly bodies are not subject to change and move in circles. Aristotle further stated that beyond Earth's sphere lies a fifth basic element, which he called the aether (αιθηρ)—a pure form of air that is not subject to change. Finally, Aristotle declared that he could analyze all material things in terms of their matter and their form (essence). Aristotle's ideas about the nature of matter and the structure of the universe dominated thinking in Europe for centuries until they were finally displaced during the Scientific Revolution.

This 1961 stamp from Greece honors the early Greek philosopher Democritus (ca. 460–ca. 370 B.C.E.). *(Author)*

Other ancient Greeks did more than contemplate the nature of matter. Through trial and error, they learned how to manipulate and use matter in a variety of intriguing ways. The Greek inventor and engineer Ctesibius of Alexandria (ca. 285–22 B.C.E.) published *On Pneumatics* in approximately 255 B.C.E. In this work, he discussed the elasticity of the air and suggested many applications of compressed air in such devices as pumps, musical instruments, and even an air-powered cannon. Some 15 years later, Ctesibus introduced a greatly improved clepsydra (water clock), which became the best timepiece in antiquity and remained unrivaled in accuracy until pendulum clocks appeared in Europe in the 17th century.

Archimedes of Syracuse (ca. 287–12 B.C.E.) was the greatest engineer of antiquity. In approximately 250 B.C.E., he designed an endless screw, later called the Archimedes screw. This important fluid-moving device could efficiently remove water from the hold of a large sailing ship as well as irrigate arid fields. Approximately 15 years later, Archimedes experienced his legendary "eureka moment" while taking a bath. As the water overflowed out of the tub, he discovered the principle of buoyancy. In his enthusiasm, he ran naked through the streets of Syracuse in Sicily to the palace of King Hieron II. Once there, he told the king that he had solved the perplexing problem of how to determine the gold content of the king's new crown without destroying it. Scientists and engineers now call this important discovery in fluid mechanics the *Archimedes principle*.

ALCHEMY

Alchemy is a mixture of mystical rituals, magical incantations, early attempts at materials science, and experimentation involving chemical reactions and procedures. Despite often haphazard practices, medieval alchemy gave rise to the science of chemistry in western Europe. The word *alchemy* comes from the Arabic word *al-kimiya* (itself derived from the late Hellenistic Greek language), meaning the "art of transmutation" as practiced by the Egyptians.

Approximately 2,500 years ago, people in many early civilizations were practicing alchemy. The form of alchemy prevalent in medieval Europe traces its ancestry back to ancient Egypt and ancient Greece. Starting in the seventh century C.E., the often mystical practices of alchemy that appeared in the Hellenistic civilization of Greece (300 B.C.E. to 300 C.E.) were absorbed and embellished by Muslim cultures around the Mediterranean Sea. Over the next several centuries, alchemy diffused into Europe.

Medieval practitioners began performing experiments and developing processes that prepared the way for the emergence of modern chemistry in western Europe. Medieval alchemists had no real understanding of the true nature of matter to guide the vast majority of their efforts. Any useful chemical reactions they uncovered were stumbled upon through tedious trial and error procedures. They usually enshrouded their procedures in puzzling and deliberately secretive jargon.

Early alchemists were influenced in their thinking by the four classic elements of ancient Greece: earth, air, water, and fire. Hellenistic-era peoples believed that all matter consisted of these four elements mixed in different proportions, so, to create different substances, the early alchemists thought that by changing the proportions of these basic elements they could change, or transmute, substances. Influenced by alchemical practices from ancient Egypt, Hellenistic Greece, and the Arab world, medieval alchemists pursued two major objectives. First, they searched for something they called the *philosopher's stone,* an object that could turn base metals into gold. Second, they pursued the *elixir of life,* a special potion that could confer immortality. Possibly influenced by Arab alchemists, who created talismans for healing, some medieval alchemists also pursued the *panacea,* a special substance that would cure all human illnesses. Finally, a few Arab and European alchemists, perhaps stimulated by the mystery of Greek fire, strove to find the *alkahest,* a presumed universal solvent that would dissolve (melt) anything and be very useful in transmutation experiments and also in warfare.

Starting in the seventh century C.E., the Byzantine emperors were able to defend the city of Constantinople (Byzantium) from attacks by invading Arab (Saracen) fleets and armies using Greek fire. Thought to have been a composition of quicklime, naphtha, and sulfur, the Byzantines used prow-mounted brass nozzles to squirt this flammable liquid onto enemy ships with devastating effects. Similar devices sprayed Greek fire from the walls of Constantinople onto attacking enemy soldiers. Though the precise formula remains a mystery, Greek fire was one of the most terrifying weapons of the medieval world and is probably responsible for the relative longevity of the Eastern Roman Empire.

In about 250 C.E., the Greek alchemist and mystic Zosimos of Panoplis (Egypt) wrote the oldest known work describing alchemy. Little is known about his personal life except for the fact that he lived in Egypt at the end of the Hellenistic period. Collaborating with other individuals, Zosimos developed an encyclopedialike collection of information about Hellenistic alchemy. His surviving works reflect not only the classical Greek notions of the four basic elements but also suggest an influence exerted by ancient Egyptian mummification practices and burial rituals.

The Arabian alchemist Abu Musa Jabir Ibn Hayyan (ca. 721–815), whose Latinized name was Gerber, is an example of the Muslim scholars—alchemists, astrologers, mathematicians, and natural philosophers—who rescued, preserved, and then transmitted the works of the ancient Greeks while western Europe languished in the Dark Ages. Among Jabir's (Gerber's) important contributions to alchemy was his modification of the four Greek classical elements to include two other special substances, sulfur (S) and mercury (Hg). He related sulfur to an idealized concept of combustion and regarded mercury as the embodiment of metallic properties. He also suggested that, with an appropriate combination of mercury and sulfur, any metal might be formed. Jabir's work encouraged many medieval alchemists to pursue the philosopher's stone in their futile attempts to turn base metals into gold.

Jabir recorded most of his experiments in an encoded form of alchemical jargon that still baffles science historians. Because these writings are so difficult to interpret, English-language scholars attribute the origin of the term *gibberish* to them. If his writings remain difficult to decipher, his contributions to the emergence of chemistry as a science are not. Jabir is generally credited with introducing an experimental approach to the practice of alchemy. He is also believed to have been the first person to synthesize hydrochloric acid (HCl) and nitric acid (HNO$_3$). Many modern chemical processes, such as distillation, crystallization, and evaporation, also trace their origins to his alchemical pursuits. Collectively, Jabir and the other notable Arab alchemists formed an effective bridge between the ancient Greek philosophers, who first pondered the nature of matter, and the western European alchemists and (later) chemists, who refined the study of chemical reactions into a science.

While pursuing various interests in alchemy in about 1247, the British independent scholar and Franciscan monk Roger Bacon (ca. 1214–92) wrote the formula for gunpowder in his encyclopedic work *Opus Majus*. Bacon recorded the following information: "Of saltpeter take 7 parts, 5 of young hazel twigs, and 5 of sulfur; and so thou wilt call up thunder and destruction, if thou know the art." The "art" Bacon referred to is alchemy. Science historians are uncertain whether he came up with the formula himself or was just acknowledging the arrival of gunpowder in Europe by way of Arab alchemists. A contemporary of Bacon, the German theologian and natural philosopher Albertus Magnus (ca. 1200–80), also practiced alchemy. He receives credit for isolating the element arsenic (As) sometime around 1250.

The Swiss physician and alchemist Paracelsus (1493–1541) trained as a medieval physician and then focused his energy on using chemicals to achieve cures. His activities helped promote medical alchemy. Paracelsus enjoyed alienating civil and church authorities and spent most of his life as a controversial figure. His abrasive behavior forced him to spend his twilight years as a wandering lecturer and intellectual renegade. Like many of his contemporaries, this medieval alchemist accepted the four classical Greek elements and the special role of mercury and sulfur prevalent in Arab alchemy, but as a Renaissance-era physician, he pioneered medical alchemy and paved the way for the role of modern chemistry in developing medicines to treat diseases. The term *iatrochemistry* refers to the combination of chemistry with medicine. The teachings of Paracelsus stressed the use of chemicals in the treatment of diseases and dominated the iatrochemical school of medicine, which flourished in western Europe from about 1525 to 1660.

Drawing heavily upon contemporary alchemical activities with metals, the German metallurgist Lazarus Ercker (ca. 1530–94) wrote "Treatise on Ores and Assaying" in 1574, a work that remained a standard reference on metallurgy and assaying for the next two centuries in Europe. In 1597, the German medical doctor and alchemist Andreas Libavius (1555–1616) published *Alchemia*—a summary of alchemy as practiced in medieval Europe. He was a student of Paracelsus and, like his mentor, emphasized the medical applications of alchemy. Some science historians regard his publication as the first chemistry textbook.

In 1667, the German alchemist Johann Joachim Becher (1635–82) introduced a refined classification of substances (especially minerals) based on his own three-element theory of matter. His work forms the basis of an early (but incorrect) theory of heat called *phlogiston theory,* which was later espoused by Georg Ernst Stahl (1660–1734) and other 18th-century scientists.

The German alchemist Hennig Brand (ca. 1630–1710) discovered the element phosphorous (P) in 1669 while searching for ways to transmute base metals into gold. Following alchemical practices of the time, Brand was heating concentrated urine in a retort when a snow-white substance appeared at the bottom of the vessel and then burst into flames. The urine evaporation method of producing phosphorous continued in Europe for

This manuscript illustration shows a medical alchemist in medieval Europe examining a flask of liquid, possibly urine. *(National Institutes of Health)*

another century, until the Swedish pharmaceutical chemist Carl Wilhelm Scheele (1742–86) prepared the element from bones (which contain calcium phosphate).

Despite the mysticism, rituals, and secrecy often found in alchemy, the individual practitioners managed to contribute to the foundations of modern chemical science. Some alchemists, such as Paracelsus, were more magician and self-promoter; others, such as Albert Magnus and Roger Bacon, were more intellectually inclined to unravel the mysteries of nature. As a result of numerous time-consuming laboratory experiments, they were able to collectively examine most available materials and stumbled upon many interesting chemical reactions. Alchemists also developed several techniques still used in modern chemistry, including distillation, fermentation, sublimation, and solution preparation.

Nevertheless, without a good understanding of the microscopic nature of matter, most alchemists became frustrated with the unexplainable results of their experiments. Many resorted to spells, incantations, magic, and secrecy to compensate for this lack of scientific understanding. Other alchemists were simply charlatans who tried to dupe gullible clients. Because of their secrecy and occasional chicanery, European alchemists often became accused of demonic activities and the practice of witchcraft. The net result of such accusations was condemnation for heresy, imprisonment, and even execution in various parts of Europe.

For all their mystical and secretive languages and rituals, alchemists did leave behind an important record of achievement with respect to how humans perceived matter. Science historians generally credit alchemists with the discovery and/or initial identification of the following five elements: antimony (Sb), arsenic (As), bismuth (Bi), phosphorous (P), and zinc (Zn). They also developed improved procedures for metal smelting, alloying, distillation, crystallization, and calcination.

As the Scientific Revolution took hold in the 17th century, it produced an important transition during which the secret activities of alchemists gave way to the openly published and freely discussed materials science activities of mechanical philosophers and pneumatic chemists—pioneering individuals who began exploring matter within the framework of the scientific method. (See chapter 4.) Leading this new wave of intellectual inquiry was the British-Irish physicist and chemist Robert Boyle, who published *The Skeptical Chymist* in 1661. In this book, Boyle attacked the tradition of the four Greek classical elements and exposed the fallacious belief of alchemists in the mythical philosopher's stone.

This was an interesting transitional period, because many of the individuals who helped establish modern science also dabbled in alchemy. For example, the British physicist and mathematician Sir Isaac Newton, whose work helped define the mechanical universe, also maintained a deep (but unpublished) interest in alchemy. In the end, individuals such as the French chemist Antoine Lavoisier (1743–94) organized the rather chaotic world of alchemy into the science of chemistry.

STEAM AND THE INDUSTRIAL REVOLUTION

Historians define the *First Industrial Revolution* as the period of enormous cultural, technical, and socioeconomic transformations that occurred in the late 18th and early 19th centuries. It served as a major stimulus for improvements in materials science and the production of advanced machines. Starting in Great Britain and quickly spreading throughout western Europe and North America, factory-based production and machine-dominated manufacturing began displacing economies based on manual labor. Historians often treat this transformation as being comparable to the Neolithic Revolution with respect to the consequences exerted on the trajectory of human civilization. Because of the global political influence of the British Empire in the 19th century, this wave of social and technical change eventually spread throughout the world.

British engineers and business entrepreneurs led the charge in the late 18th century by developing steam power (fueled by coal) and using powered machinery in manufacturing (primarily in the textile industry). These technical innovations were soon followed by the development of all-metal machine tools in the early 19th century. The availability of these machine tools quickly led to the development of more advanced machines for use in factory-based manufacturing in other industries. More and less expensive goods became available. Workers left the farms and flocked to urban areas to work in factories.

A common misconception is that the Scottish engineer James Watt (1736–1819) invented the steam engine. Actually, Watt was working as an instrument maker at the University of Glasgow in 1764 when officials at that institution gave him a Newcomen engine to repair. The repair project introduced Watt to the world of steam, and the inventive young man soon not only repaired the Newcomen engine but greatly improved its design. Watt's stroke of genius was a decision to add a separate condenser to the heat engine, thereby greatly improving its thermodynamic efficiency. The British engineer Thomas Newcomen (1663–1729) had developed and patented the original Newcomen engine at the start of the 18th century. The primary application of this bulky and thermally inefficient heat engine was to operate as a steam-powered vacuum engine that could drain water out of deep mines.

In 1775, Watt entered a business partnership with the British entrepreneur Matthew Boulton (1728–1809). Continually being improved, the Watt steam engine soon dominated mechanical power production all over the United Kingdom and eventually throughout much of continental Europe and North America. Steam provided power for factories, which no longer had to be located near sources of falling water (hydropower). Other versions of the Watt steam engine began to power ships and propel railroad trains. In a few short years, steam-powered engines transformed human civilization.

As scientists and engineers sought to make steam engines more efficient, they began to explore the nature of heat and its relationship to mechanical work. They also investigated the thermal properties of matter, giving rise to the science of thermodynamics. Steam not only powered the First Industrial Revolution, it also powered an amazing intellectual revolution in the 19th century involving materials science, fluid mechanics, thermodynamics, and the rebirth of atomic theory.

CHAPTER 4

Quantifying Matter during the Scientific Revolution

The ability of human beings to relate the microscopic (atomic level) behavior of matter to readily observable macroscopic properties (such as density, pressure, and temperature) has transformed the world. The initial breakthroughs in accurately quantifying matter took place during the Scientific Revolution. This chapter explores the pioneering efforts of some of the individual geniuses whose discoveries established the foundations of modern science. The scientific method emerged as an important new way of viewing the physical world and continues to exert a dominant influence on human civilization.

AN OVERVIEW OF THE SCIENTIFIC REVOLUTION

The Scientific Revolution was the intellectually tumultuous period between 1543 and 1687 in western Europe, when many creative people developed important new ways of looking at matter and physical phenomena. In 1543, Nicholas Copernicus challenged the long-held geocentric cosmology of Aristotle, who advocated that Earth was the center of the universe. Copernicus's book *On the Revolution of Celestial Spheres* promoted a radically new perspective called heliocentric cosmology. Early in the 17th century, Galileo Galilei used a telescope to revolutionize astronomy and soon vigorously endorsed heliocentric cosmology. (Scientists commonly refer to Galileo Galilei by just his first name.) Galileo also performed a variety of important observations and experiments that established the investigative approach known as the scientific method. Sir Isaac Newton created the intellectual capstone of the Scientific Revolution when he published his magnificent work *Mathematical Principles of Natural Philosophy* (short title *The Principia*) in 1687. This monumental work transformed the practice of physical science and completed the revolution started by Copernicus.

The achievements of Galileo and Newton supported deeper exploration into the nature of matter and the operation of the universe by other

THE SCIENTIFIC METHOD

Science provides an organized way of learning about the physical universe and the fundamental principles that govern the relationships between energy and matter. The scientific method is the self-correcting, iterative process by which people acquire new knowledge about the universe. There are several general steps in the scientific method.

Stimulated by curiosity, a scientist might start the process by making a hypothesis (or educated guess) about some natural phenomenon or process. He (or she) then performs a series of reproducible experiments or observations to gather data that lend support to the hypothesis. Should the experimental data or observations suggest an interesting trend or relationship among physical properties, the scientist might create a predictive analytical model to better characterize the phenomenon or process being studied. The scientist then validates this model with additional experiments or reproducible observations.

At this point, the scientist usually communicates with other scientists and invites them to evaluate the findings. If supported by many additional experiments or reproducible observations, the scientists might collectively decide to elevate the new hypothesis or model to the status of a physical law or theory.

All credible scientists understand that any hypothesis, model, physical law, or even cherished theory is always subject to challenge and renewed scrutiny as part of the scientific method. Additional testing or more precise observations may provide results that cause them to refine, expand, or perhaps even discard a previously accepted hypothesis, model, or law. Tossing out old concepts in the face of new, improved data represents the inherent self-correcting nature of the scientific method.

scientists. As discussed later in this chapter, their influence continued well into the 18th and 19th centuries. Some scientists, such as the Swiss mathematician Daniel Bernoulli (1700–82), incorporated Galileo's work in hydraulics, Newton's laws of motions, the notion of the indestructibility of matter, and the conservation of energy principle to develop the important field of *fluid mechanics*. Scientists define fluid mechanics as the branch of classical mechanics that deals with fluid motions and with the forces exerted on solid bodies in contact with fluids. The term *fluid* encompasses both liquids and gases.

While carefully investigating the process of combustion and associated chemical reactions, the French scientist Antoine-Laurent Lavoisier created the framework of modern chemistry. Other scientists such as Robert Boyle and Joseph Louis Gay-Lussac (1778–1850) experimented with different gases—including and especially air. The combined efforts of these so-called pneumatic chemists spanned more than a century of research and ultimately resulted in an important physical law relating the temperature, pressure, and volume of gaseous substances. Scientists call this principle the ideal gas equation of state.

The phenomenon of heat has puzzled human beings since prehistoric times. Starting in the 17th century, European alchemists and chemists initially embraced phlogiston theory in their attempts to explain combustion. In the 18th century, Lavoisier used precise mass-balances in conjunction with chemical combustion experiments to displace that erroneous theory. Then, however, he and his contemporaries supported the equally wrong concept of heat known as the caloric theory. It required meticulous research efforts by scientists in the 19th century to overthrow caloric theory and firmly establish the modern concept of heat as energy in transit due to a temperature difference.

In the early 1840s, the British physicist James Prescott Joule (1818–89) used his own private laboratory to quantitatively investigate the nature of heat. He discovered the very important relationship between heat and mechanical work. (The concept of work is discussed in the next section.) Although initially met with resistance, Joule's discovery of the mechanical equivalence of heat was eventually recognized as a meticulous experimental demonstration of the conservation of energy principle—one of the intellectual pillars of thermodynamics.

Scientists and engineers now understand the physical mechanisms (both microscopic and macroscopic) that underlie the three basic modes of heat transfer: conduction, convection, and radiation. Conduction heat

A fire whirl in a modern research facility. The detailed study of combustion by 17th- and 18th-century scientists led to many important discoveries in chemistry. *(National Institute of Standards and Technology)*

transfer takes place because of the motion of individual atoms or molecules. Convection involves the transfer of heat by bulk fluid motion. Radiation involves the transfer of energy by electromagnetic waves. (See chapter 5.) Normally associated with the phenomenon of radiation heat transfer is infrared radiation, which is that portion of the electromagnetic spectrum that lies between visible light and radio waves. The science of thermodynamics and the principles of engineering heat transfer provide the technical framework for today's energy-dependent global civilization. None of these important technical developments would have been possible without the emergence of the scientific method during the Scientific Revolution.

THE MECHANICS OF THE UNIVERSE

As part of the Scientific Revolution, Galileo and Newton provided the first useful quantitative interpretation of the relationships between matter, motion, energy, and gravity. This section presents the classical interpretation of one of nature's most familiar yet mysterious phenomena—gravity. Einstein extended their trailblazing work on gravity and the motion of celestial objects when he proposed general relativity early in the 20th century.

Gravity is the attractive force that tugs on people and holds them on the surface of Earth. Gravity also keeps the planets in orbit around the Sun and causes the formation of stars, planets, and galaxies. Where there is matter, there is gravity. In a simple definition, gravity is the pull material objects exert on one another. A baseball thrown up into the air will rise to a certain height, pause at the apex of its trajectory, and then begin its downward trip to Earth under gravity's influence.

People most often remember Galileo as the first astronomer to use a telescope to view the heavens and to conduct the astronomical observations that supported the Scientific Revolution. However, Galileo was also the physicist who founded the science of mechanics and provided Newton the underlying data and ideas that allowed him to develop the laws of motion and the universal law of gravitation.

In the late 16th century, European professors taught natural philosophy (physics) as metaphysics, an extension of Aristotelian philosophy. Before Galileo's pioneering contributions, physics was not considered an experimental science. His research activities successfully challenged the 2,000-year tradition of learning based on ancient Greek philosophy.

Through his skillful use of mathematics and innovative experiments, Galileo helped establish the important investigative technique called the scientific method.

Aristotle had stated that heavy objects fall faster than lighter objects. Galileo disagreed and held the opposite view—namely, that except for air resistance, two objects would fall at the same rate regardless of their masses. It is not certain whether he personally performed the legendary musket ball versus cannon ball drop experiment from the Leaning Tower of Pisa to prove this point. He did, however, conduct a suffi-

BLIND ALLEYS: PHLOGISTON AND CALORIC THEORY

In the 17th and 18th centuries, scientists used phlogiston theory to explain combustion and the process of oxide formation. According to this now-discarded theory of heat, all flammable materials contained phlogiston, which was regarded as a weightless, colorless, and tasteless substance. During combustion, the flammable material released its phlogiston, and what remained of the original material was a dephlogisticated substance—a crumby or calcined residual material called *calx.* Phlogistonists assumed that when the phlogiston left a substance during combustion, it entered the air, which then became "phlogisticated air." Their pioneering combustion experiments with air in sealed containers revealed some very interesting yet initially inexplicable results.

The French chemist Antoine Lavoisier performed carefully conducted mass-balance experiments in which he demonstrated that combustion involved the chemical combination of substances with oxygen. His work clearly dispelled the notion of phlogiston. In 1787, Lavoisier introduced the term *caloric* to support his own theory of heat, a theory in which heat was regarded as a weightless, colorless fluid that flowed from hot to cold. An incorrect concept called the conservation of heat principle served as one of the major premises in Lavoisier's caloric theory. Another important hypothesis of this now-discarded theory of heat was that caloric fluid consisted of tiny, self-repelling particles that flowed easily into a substance, expanding the substance as its temperature increased.

In 1824, the French engineering physicist Sadi Carnot (1746–1832) published *Reflections on the Motive Power of Fire,* in which he correctly defined the maximum thermodynamic efficiency of a heat engine using incorrect caloric theory.

cient number of experiments involving objects rolling or sliding down inclined planes to upset Aristotelian thinking and create the science of mechanics.

During his lifetime, Galileo was limited in his study of motion by an inability to accurately measure small increments of time. No one had yet developed a timekeeping device capable of accurately measuring 10ths, 100ths, or 1,000ths of a second. Despite this severe impediment, he conducted many important experiments that produced remarkable insights into the physics of free fall and projectile motion.

Since steam engines dominated the First Industrial Revolution, the thermodynamic importance of Carnot's work firmly entrenched the erroneous concept of heat throughout first half of the 19th century. It took pioneering experiments by the British-American scientist Benjamin Thomson (1753–1814) (later known as Count Rumford) and James Prescott Joule to dislodge the caloric theory of heat.

In 1798, Thomson published the important paper "An Experimental Inquiry Concerning the Source of Heat Excited by Friction." He had conducted a series of experiments linking the heat released during cannon boring operations in a Bavarian armory to mechanical work, so Thomson suggested in his paper that heat was the result of friction rather than the flow of the hypothetical fluid substance called caloric. The more Thompson bored a particular cannon barrel, the more heat appeared due to mechanical friction—an experimental result that clearly violated (and thus disproved) the conservation of heat principle of caloric theory.

Joule completed the assault on caloric theory in the mid-1840s when his series of precisely instrumented experiments linked the mechanical motion of paddlewheels stirring water to a subtle rise in temperature. Joule's pioneering research demonstrated that mechanical work and heat are equivalent. His experimental results represent one of the great breakthroughs in science.

From the late 19th century forward, scientists no longer viewed heat as an intrinsic property of matter, as was erroneously done in both phlogiston theory and caloric theory. Heat, like work, became recognized as a form of energy in transit. On a microscopic scale, it is sometimes useful to envision heat as disorganized energy in transit—that is, the somewhat chaotic processes taking place as molecules or atoms randomly experience more energetic collisions and vibrations under the influence of temperature gradients.

Galileo's experiments demonstrated that, in the absence of air resistance, all bodies, regardless of mass, fall to the surface of Earth at the same *acceleration.* Physicists define acceleration as the rate at which the velocity of an object changes with time. On or near Earth's surface, the acceleration due to gravity (symbol: *g*) of a free-falling object has the standard value of 32.1737 ft/s² (9.80665 m/s²) by international agreement. Galileo's study of projectile motion and free-falling bodies led to the incredibly important formula that helped Newton create his laws of motion and universal law of gravitation. The key formula (d = ½ g t²) mathematically describes the distance (d) traveled by an object in free fall (neglecting air resistance) in time (t).

Newton was a brilliant British physicist, mathematician, and astronomer whose law of gravitation, three laws of motion, development of the calculus, and design of a new type of reflecting telescope made him one of the greatest scientific minds in human history. Through the patient encouragement and financial support of the British mathematician Sir Edmund Halley (1656–1742), Newton published his great work, *The Principia,* in 1687. This monumental book transformed the practice of physical science and completed the Scientific Revolution started by Copernicus a century earlier. Newton's three laws of motion and universal law of gravitation still serve as the basis of classical mechanics.

In August 1684, Halley traveled to Newton's home at Woolsthorpe, England. During this visit, Halley convinced the reclusive genius to address the following puzzle about planetary motion: What type of curve does a planet describe in its orbit around the Sun, assuming an inverse square law of attraction? To Halley's delight, Newton responded: "An ellipse." Halley pressed on and asked Newton how he knew the answer to this important question. Newton nonchalantly informed Halley that he had already done the calculations years ago (in about 1666). Since the absent-minded Newton could not find his old calculations, he promised to send Halley another set as soon as he could.

To partially fulfill his promise, Newton later that year (1684) sent Halley his *De Motu Corporum (On*

This German stamp honors the British scientist Sir Isaac Newton. *(Department of Energy)*

the motion of bodies). Newton's paper demonstrated that the force of gravity between two bodies is directly proportional to the product of their masses and inversely proportional to the square of the distance between them. (Physicists now call this physical relationship Newton's universal law of gravitation.) Halley was astounded and implored Newton to carefully document all of his work on gravitation and orbital mechanics. Through the patient encouragement and financial support of Halley, Newton published *The Principia* in 1687. Many scientists regard *The Principia* as the greatest technical accomplishment of the human mind.

The science of mechanics links the motion of a material object with the measurable quantities of mass, velocity, and acceleration. Through Newton's efforts, the term *force* entered the lexicon of physics. Scientists say a force influences a material object by causing a change in its state of motion. (Historically, Newton used the term *fluction* when he developed the calculus.) The concept of force emerges out of Newton's second law of motion. In its simplest and most familiar mathematical form, force (F) is the product of an object's mass (m) and its acceleration (a)—namely, $F = m\,a$. In his honor, scientists call the fundamental unit of force in the SI system the newton (N). One newton of force accelerates a mass of one kilogram at the rate of one meter per second per second ($1\,N = 1\,kg\text{-}m/s^2$). In the American customary system, engineers define one pound-force (lbf) as the force equivalent to the weight of a one-pound mass (1 lbm) object at sea level on Earth. (Engineers find it helpful to remember that 1 N = 0.2248 lbf.)

Physicists define *energy* (E) as the ability to do work. Within classical Newtonian physics, scientists describe *mechanical work* (W) as a force (F) acting through a distance (d). The amount of work done is proportional to both the force involved and the distance over which the force is exerted in the direction of motion. A force perpendicular to the direction of motion performs no work. In the SI system, scientists measure energy with a unit called the joule (J). One joule represents a force of one newton (N) moving through a distance of one meter (m). This unit honors the British physicist James Prescott Joule. In the American customary system, engineers often express energy in terms of the British thermal unit (Btu). (One Btu equals 1,055 joules.)

Within classical physics, energy (E), work (W), and distance (d) are scalar quantities, while velocity (v), acceleration (a), and force (F) are vector quantities. A scalar is a physical quantity that has magnitude only, while a vector is a physical quantity that has both magnitude and direction at each point in space.

One of the most important contributions of Western civilization to the human race was the development of the scientific method and, through it, the start of all modern science. During the intellectually turbulent 17th century in western Europe, people of great genius began identifying important physical laws and demonstrating experimental techniques that helped humans everywhere better understand the physical universe.

CONCEPTS OF DENSITY, PRESSURE, AND TEMPERATURE

This section discusses three familiar physical properties: density, pressure, and temperature. By thinking about the atoms that make up different materials, scientists and engineers have been able to understand, quantify, and predict how interplay at the atomic (microscopic) level results in the physical properties that are measurable on a macroscopic scale. This intellectual achievement has transformed human civilization.

Density

To assist in more easily identifying and characterizing different materials, scientists devised the material property called density, one of the most useful macroscopic physical properties of matter. Solid matter is generally denser than liquid matter, and liquid matter denser than gases. Scientists define *density* as the amount of mass contained in a given volume. They frequently use the lower case Greek letter rho (ρ) as the symbol for density in technical publications and equations.

Scientists use the density (ρ) of a material to determine how massive a given volume of that particular material would be. Density is a function of both the atoms from which a material is composed as well as how closely packed the atoms are arranged in the particular material. At room temperature (nominally 68°F [20°C]) and one atmosphere pressure, the density of some interesting solid materials is as follows: gold 1,205 lbm/ft³ (19,300 kg/m³ [19.3 g/cm³]); iron 493 lbm/ft³ (7,900 kg/m³ [7.9 g/cm³]); diamond (carbon) 219 lbm/ft³ (3,500 kg/m³ [3.5 g/cm³]); aluminum 169 lbm/ft³ (2,700 kg/m³ [2.7 g/cm³]); and bone 112 lbm/ft³ (1,800 kg/m³ [1.8 g/cm³]). Like most gases at room temperature and one atmosphere pressure, oxygen has a density of just 0.083 lbm/ft³ (1.33 kg/m³ [1.33×10^{-3} g/cm³])—a value that is about 1,000 times less than the density of most solid or liquid materials normally encountered on Earth's surface.

Scientists know that the physical properties of matter are often interrelated. Namely, when one physical property (such as temperature) increases or decreases, other physical properties (such as volume or density) also change. As a direct result of the Scientific Revolution, people learned how to define the behavior of materials by developing special mathematical expressions termed *equations of state*. Scientists developed these mathematical relationships using both theory and empirical data from many carefully conducted laboratory experiments.

Pressure

Scientists describe pressure (P) as force per unit area. The most commonly encountered American customary unit of pressure is pounds-force per square inch (psi). In the SI system, the fundamental unit of pressure is the pascal (Pa) in honor of Blaise Pascal (1623–62). The 17th-century French scientist conducted many pioneering experiments in fluid mechanics. One pascal represents a force of one newton (N) exerted over an area on one square meter, that is, $1\ Pa = 1\ N/m^2$. One psi is approximately equal to 6,895 Pa.

Anyone who has plunged into the deep end of a large swimming pool and then descended to the bottom of the pool has personally experienced the phenomenon of hydrostatic pressure. Hydrostatic pressure is the pres-

OTTO VON GUERICKE'S AMAZING EXPERIMENT

In 1654, the German scientist and politician Otto von Guericke (1602–86) provided a dramatic public demonstration of atmospheric pressure in the city of Magdeburg (and later elsewhere in Germany). No stranger to science, he had invented an air pump several years earlier. His famous experiment involved two hollow metal hemispheres each about 1.64 feet (0.5 m) in diameter. By most historic accounts, once von Guericke had joined the hemispheres and pumped the air out, two teams of eight horses could not pull apart the assembled sphere, yet the hemispheres easily separated when von Guericke turned a valve and let air back into the evacuated sphere.

sure at a given depth below the surface of a static (nonmoving) fluid. As Pascal observed in the 17th century, the greater the depth, the greater the pressure.

Atmospheric pressure plays an important role in many scientific and engineering disciplines. In an effort to standardize their research activities, scientists use the following equivalent atmospheric pressure values for sea level: one atmosphere (1 atm) ≡ 760 mm of mercury (Hg) (exactly) = 29.92 in Hg = 14.695 psi = 1.01325×10^5 Pa.

ELASTICITY OF MATTER

Engineers measure the strength of materials by the capacity of substances, such as metals, concrete, wood, glass, and plastics, to withstand stress and strain. As part of the Scientific Revolution, scientists began investigating the strengths of materials in an organized, quantitative way.

The British scientist Robert Hooke (1635–1703) studied the action of springs in 1678 and reported that the extension (or compression) of an elastic material takes place in direct proportion to the force exerted on the material. Today, physicists use Hooke's law to quantify the displacement associated with the restoring force of an ideal spring.

Scientists define *elasticity* as the ability of a body to resist a stress (distorting force) and then return to its original shape when the stress is removed. There are several types of stresses: tension (due to a pulling force), compression (due to a pushing force), torsion (due to a rotating or twisting force), and shear (due to internal slipping or sliding). All solid objects are elastic (to some degree) when they experience deformation. The degree of elasticity, however, is a function of the material.

According to Hooke's law, within the elastic limit of a particular substance, the stress is proportional to the strain. If a person pulls on a metal spring (that is, exerts a tension or pulling force), the spring will experience a strain and stretch a certain amount in proportion to the applied tensile force. Within the material's elastic limit, once the person stops pulling on the spring, it restores itself to the original position. This also occurs with springs that work in compression. If the person keeps pulling on the spring, however, and stretches (deforms) it beyond the material's elastic limit, then the spring experiences a plastic deformation. Once the tension force is released after a plastic deformation, the material can no longer return to its original shape and dimensions. If the person keeps pull-

One important feature of Earth's atmosphere is that the density in a column of air above a point on the planet's surface is not constant. Density and atmospheric pressure decrease with increasing altitude until both become negligible in outer space. The pioneering work of Pascal and the Italian physicist Evangelista Torricelli (1608–47) guided other scientists in measuring and characterizing Earth's atmosphere.

Engineers often treat rigid, solid bodies as incompressible objects. Generally, very high values of force are needed to compress or deform a rigid,

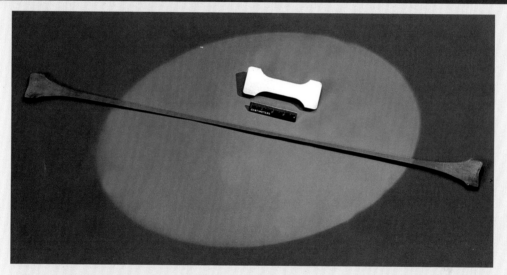

In the spotlight here is a "superplastic" ultrahigh-carbon steel that can be stretched (plastically deformed) to elongations exceeding 1,100 percent of the original size. Such superplastic metals can dramatically reduce the need for machining parts. *(Department of Energy)*

ing on the metal spring after plastic deformation, the metal stretches to the yield point, at which it breaks or fractures.

The field of solid mechanics deals with the behavior of solid materials under a variety of external influences, including stress forces. Some materials, such as steel, are much better at resisting tensile (pulling) forces; other materials, such as concrete, are much better at resisting compressive (pushing) forces. The strengths of various materials are ultimately determined by the arrangement of the atoms and molecules that make up the materials.

solid body. Unlike rigid solids, fluids are materials that can flow, so engineers use pressure differentials to move fluids. They design pumps to move liquids (often treated as incompressible fluids), while they design fans to move gases (compressible fluids). An incompressible fluid is assumed to have a constant value of density; a compressible fluid has a variable density. One of the interesting characteristics of gases is that, unlike solids or liquids, they can be compressed into smaller and smaller volumes.

Temperature

While temperature is one of the most familiar physical variables, it is also one of the most difficult to quantify. Scientists suggest that on the macroscopic scale temperature is the physical quantity that indicates how hot or cold an object is relative to an agreed upon standard value. Temperature defines the natural direction in which energy flows as heat—namely, from a higher temperature (hot) region to a lower temperature (cold) region. Taking a microscopic perspective, temperature indicates the speed at which the atoms and molecules of a substance are moving.

Scientists recognize that every object has the physical property called temperature. They further understand that when two bodies are in thermal equilibrium, their temperatures are equal. A thermometer is an instrument that measures temperatures relative to some reference value. As part of the Scientific Revolution, creative individuals began using a variety of physical principles, natural references, and scales in their attempts to quantify the property of temperature.

In about 1592, Galileo attempted to measure temperature with a device he called the thermoscope. Although Galileo's work represented the first serious attempt to harness the notion of temperature as a useful scientific property, his innovative thermoscope did not supply scientifically significant temperature data.

The German physicist Daniel Gabriel Fahrenheit (1686–1736) was the first to develop a thermometer capable of making accurate, reproducible measurements of temperature. In 1709, he observed that alcohol expanded when heated and constructed the first closed-bulb glass thermometer, with alcohol as the temperature-sensitive working fluid. Five years later (in 1714), he used mercury as the thermometer's working fluid. Fahrenheit selected an interesting three-point temperature reference scale for his original thermometers. His zero point (0°F) was the lowest temperature he could achieve with a chilling mixture of ice, water, and ammonium chloride (NH_4Cl). Fahrenheit then used a mixture of just water and ice as his second reference temperature (32°F). Finally, he chose his own body temperature (recorded as 96°F) as the scale's third reference temperature.

After his death, other scientists revised and refined the original Fahrenheit temperature scale, making sure there were 180 degrees between the freezing point of water (32°F) and the boiling point of water (212°F) at one atmosphere pressure. On this refined scale, the average temperature of the human body appeared as 98.6°F. Although the Fahrenheit temperature scale is still used in the United States, most other nations have adopted another relative temperature scale, called the Celsius scale.

In 1742, the Swedish astronomer Anders Celsius (1701–44) introduced the relative temperature scale that now carries his name. He initially selected the upper (100-degree) reference temperature on his new scale as the freezing point of water and the lower (0-degree) reference temperature as the boiling of water at one atmosphere pressure. He then divided the scale into 100 units. After Celsius's death, the Swedish botanist and zoologist Carl Linnaeus (1707–78) introduced the present-day Celsius scale thermometer by reversing the reference temperatures. The modern Celsius temperature scale is a relative temperature scale in which the range between two reference points (the freezing point of water at 0°C and the boiling point of water at 100°C) are conveniently divided into 100 equal units, or degrees.

Scientists describe a *relative temperature scale* as one that measures how far above or below a particular temperature measurement is with respect to a specific reference point. The individual degrees, or units, in a relative scale are determined by dividing the relative scale between two known reference temperature points (such as the freezing and boiling points of water at one atmosphere pressure) into a convenient number of temperature units (such as 100 or 180).

Despite considerable progress in thermometry in the 18th century, scientists still needed a more comprehensive temperature scale, namely, one that included the concept of absolute zero, the lowest possible temperature, at which molecular motion ceases. The Irish-born British physicist William Thomson Kelvin (1824–1907), first baron of Largs, proposed an absolute temperature scale in 1848. The scientific community quickly embraced Kelvin's temperature scale. The proper SI term for temperature is *kelvins* (without the use of *degree*), and the proper symbol is K (without the symbol °). Scientists generally use absolute temperatures in such disciplines as physics, astronomy, and chemistry; engineers use either relative or absolute temperatures in thermodynamics, heat transfer analyses, and

RANKINE—THE OTHER ABSOLUTE TEMPERATURE

Most of the world's scientists and engineers use the Kelvin scale to express absolute thermodynamic temperatures, but there is another absolute temperature scale called the Rankine scale (symbol R) that sometimes appears in engineering analyses performed in the United States—analyses based upon the American customary system of units. In 1859, the Scottish engineer William John Macquorn Rankine (1820–72) introduced the absolute temperature scale that now carries his name. Absolute zero in the Rankine temperature scale (that is, 0 R) corresponds to –459.67°F. The relationship between temperatures expressed in rankines (R) and those expressed in degrees Fahrenheit (°F) is T (R) = T (°F) + 459.67. The relationship between the Kelvin scale and the Rankine scale is (9/5) × absolute temperature (kelvins) = absolute temperature (rankines). For example, a temperature of 100 K is expressed as 180 R. The use of absolute temperatures is very important in science.

mechanics, depending on the nature of the problem. Absolute temperature values are always positive, but relative temperatures can have positive or negative values.

LAVOISIER AND THE RISE OF MODERN CHEMISTRY

Antoine-Laurent Lavoisier was the great French scientist who founded modern chemistry. Although trained as a lawyer, his interest in science soon dominated his lifelong pursuits. Born into wealth, this multitalented individual conducted scientific research from the 1760s onward while also engaged in a variety of political, civic, and financial activities.

In 1768, Lavoisier became a member of Ferme Générale, a private, profit-making organization that collected taxes for the royal government. Although this financially lucrative position opened up other political opportunities for him, an association with tax collecting for the king would prove fatal. Lavoisier began an extensive series of combustion experiments in 1772, the primary purpose of which was to overthrow phlogiston theory. His careful analysis of combustion products allowed Lavoisier to demonstrate that carbon dioxide (CO_2) formed. He advocated the caloric theory of heat and suggested that combustion was a process in which the burning substance combined with a constituent of the air—a gas he called *oxygine* (a term meaning acid maker).

As part of his political activities, Lavoisier received an appointment in 1775 to serve as one of the commissioners at the arsenal in Paris. He became responsible for the production of gunpowder. His official duties encouraged him to construct an excellent laboratory at the arsenal and to establish professional contacts with scientists throughout Europe.

One of Lavoisier's contributions to the establishment of modern chemistry was his creation of a system of chemical nomenclature, which he summarized in the 1787 publication *Method of Chemical Nomenclature*. Lavoisier's book promoted the principle by which chemists could assign every substance a specific name based on the elements of which it was composed. His nomenclature system allowed scientists throughout Europe to quickly communicate and compare their research, thus enabling a rapid advance in chemical science during that period. Modern chemists still use Lavoisier's nomenclature.

In 1789, he published *Treatise of Elementary Chemistry,* which science historians regard as the first modern textbook in chemistry. In the book, Lavoisier discredited the phlogiston theory of heat and championed caloric theory. He advocated the important law of conservation of mass and suggested that an element was a substance that could not be further broken down by chemical means. Lavoisier also provided a list of all known elements in this book—including light and caloric (heat), which he erroneously regarded as fluidlike elemental substances.

Despite nurturing great scientific achievements, these were troubled times in France. When the French Revolution took place in 1789, Lavoisier initially supported it because he had previously campaigned for social reform. Throughout his life, he remained an honest though politically naïve individual. Lavoisier's privileged upbringing had socially insulated him from appreciating the growing restlessness of the peasants. He did not properly heed the growing dangers around him and flee France. When the revolution became more extreme and transitioned into the Reign of Terror (1792–94), his membership in the much-hated royal tax collecting organization (Ferme Générale) made him an obvious target for arrest and execution. All his brilliant contributions to the science of chemistry could not save his life. After a farcical trial, he lost his head to the guillotine in Paris on May 8, 1794.

DALTON REVIVES INTEREST IN THE ATOMIC NATURE OF MATTER

In 1649, the French scientist and philosopher Pierre Gassendi (1592–1655) revisited the atomic theory of matter. His efforts served as an important bridge between ancient Greek atomism and the subsequent revival of atomic theory in the 19th century.

Science historians generally credit the British schoolteacher and chemist John Dalton with the revival of atomism and the start of modern atomic theory. The first step in Dalton's efforts occurred in 1801, when he observed that the total pressure of a mixture of gases is equal to the sum of the par-

AVOGADRO'S BOLD HYPOTHESIS

In 1808, Joseph Gay-Lussac performed several gas reaction experiments. Examining the results of his research, the French scientist developed the law of combining volumes. This law states that at the same pressure and temperature, the volumes of reactant gases are in the ratios of whole numbers.

Three years later (in 1811), Amedeo Avogadro reviewed Gay-Lussac's work and developed his own interpretation of the law of combining volumes. The Italian scientist hypothesized that equal volumes of any two gases at the same temperature and pressure contain the same number of atoms or molecules. Fellow scientists shunned Avogadro's bold hypothesis for many years, possibly because no one at the time had a clear understanding of the true nature of atoms or molecules.

Scientists did revisit Avogadro's visionary hypothesis and then rephrased it in terms of the SI unit of substance called the mole (mol). By international agreement (since 1971), scientists define a mole as the amount of substance that contains as many elementary units as there are atoms in 0.012 kg (12 g) of the isotope carbon-12. The elementary units in the definition of the mole may be specified as atoms, molecules, ions, or radicals. Twelve grams of carbon-12 (that is, one mole) contains exactly 6.02×10^{23} atoms. This precise number of atoms represents a quantity called Avogadro's number (N_A). Scientists treat Avogadro's number as a physical constant with the value 6.02×10^{23} mol^{-1}. While Avogadro never personally attempted to measure the value of the constant that now carries his name, his brilliant insight into the nature of matter played an important role in the development of modern science.

tial pressures of each individual gas that makes up the mixture. Scientists now call this scientific principle Dalton's law of partial pressures.

Dalton's interest in the behavior of gases allowed him to quantify the atomic concept of matter. In 1803, he suggested that each chemical element was composed of a particular type of atom. He defined the atom as the smallest particle (unit) of matter in which a particular element can exist. He also showed how the relative masses, or weights, of different atoms could be determined. To establish his relative scale, he assigned the atom of hydrogen a mass of unity. In so doing, Dalton revived atomic theory and inserted atomism into the mainstream of modern science.

In 1811, the Italian scientist Amedeo Avogadro formulated a famous hypothesis that eventually became known as Avogadro's law. He proposed that equal volumes of gases at the same temperature and pressure contain equal numbers of molecules. At the time, scientists did not have a clear understanding of the difference between an atom and a molecule.

Later in the 19th century, scientists recognized the molecule as the smallest particle of any substance (element or compound) as it normally occurs. By the time the Russian scientist Dmitri Mendeleev published his famous periodic law in 1869, it was generally appreciated within the scientific community that molecules, such as water (H_2O), consisted of collections of atoms.

Throughout the remainder of the 19th century, scientists generally remained comfortable with the basic assumption that the atoms of a chemical element were simply very tiny, indivisible spheres. Scientists such as the Austrian theoretical physicist Ludwig Boltzmann (1844–1906) used the solid atom model to develop important new scientific concepts, such as the kinetic theory of gases and the principles of statistical thermodynamics. Few scientists dared to speculate that the tiny solid atom might actually have an internal structure—one characterized by much smaller subatomic particles and a vast quantity of empty space.

The wake-up call that forced the development of a revised model of the atom came in 1897, when the British physicist Sir Joseph J. Thomson (1856–1940) discovered the electron—the first known subatomic particle. Upcoming chapters describe how scientists developed the nuclear atom model.

CHAPTER 5

Understanding Matter's Electromagnetic Properties

Τhis chapter discusses electromagnetism, one of the more interesting and important properties of matter. From telecommunications to computers, from electronic money transfers to home entertainment centers, the application of electromagnetic phenomena dominates contemporary civilization. People everywhere are immersed in a world controlled by the manipulation and flow of electrons.

MAGNETISM AND THE MYSTERIES OF THE LODESTONE

The history of magnetism extends back to antiquity. Historic tradition suggests that the Greek playwright Euripides (ca. 480–06 B.C.E.) applied the word *magnes* (μαγνης) to a type of ore—now called magnetite—that displayed a preferential attraction for iron. The English word *magnet* traces its origin to the ancient Greek expression "stone of magnesia" (μαγνητης λιθος). The reference is to an interesting type of iron ore named *lodestone* by alchemists that is found in Magnesia, a region in east central Greece. Scientists also named the chemical elements magnesium (Mg) and manganese (Mn) for substances associated with this geographic location.

Lodestone is an intensely magnetic form of magnetite—a mineral consisting of iron oxide (Fe_3O_4). Although magnetite is a commonly found mineral, lodestone is rare. Contemporary scientific research suggests that not all pieces of magnetite can become a lodestone; a certain crystal structure and composition that experiences a strong transient magnetic field is required. Nature provides this transient magnetic field during a lightning strike. Lightning involves a large electric current lasting just a fraction of a second, during which time the enormous electric charge also produces a strong magnetic field. Experiments have demonstrated how lightning strikes could transform mineral samples of magnetite with the appropriate crystalline structure into lodestones.

Until the Chinese used lodestone to develop an early mariner's compass in the late 11th century, little progress was made by humans in either understanding or applying the phenomenon of natural magnetism. The notion of a compass, consisting of a slender sliver of lodestone suspended on a string, appeared in Europe about a century later. The compass is a device that uses the intrinsic properties of a magnetized metal pointer (such as lodestone) to indicate the general direction of Earth's magnetic North Pole. Earth is actually a gigantic magnet, with one pole geographically located in northern Canada (called the north magnetic pole) and the other pole in Antarctica (the south magnetic pole). The North Pole–pointing mariner's compass improved the art of open water navigation and greatly facilitated global exploration by sailing ships of European nations in the 15th and 16th centuries.

In 1600, the British physician and scientist William Gilbert (1544–1603) published *De Magnete* (On the magnet), the first technical book on the subject. His pioneering study distinguished between electrostatic and magnetic effects. Gilbert received an appointment as the royal physician to Great Britain's Queen Elizabeth I in 1601. He was also an excellent scientist who explored all previous observations concerning the magnetic and electrostatic properties of matter. As part of the ongoing Scientific Revolution, Gilbert systematically investigated and extended these ancient observations, frequently enhancing them with clever experiments of his own construction.

Testing many different materials, he listed all that exhibited the same electrostatic property as amber. In so doing, he introduced the Latin word *electricus* to describe an amberlike attractive property. The ancient Greek word for amber is *electron* (ηλεκτρον), from which the modern English words *electron* and *electricity* derive. Gilbert also examined natural magnetism very thoroughly and demonstrated that he could magnetize steel rods by stroking them with lodestone.

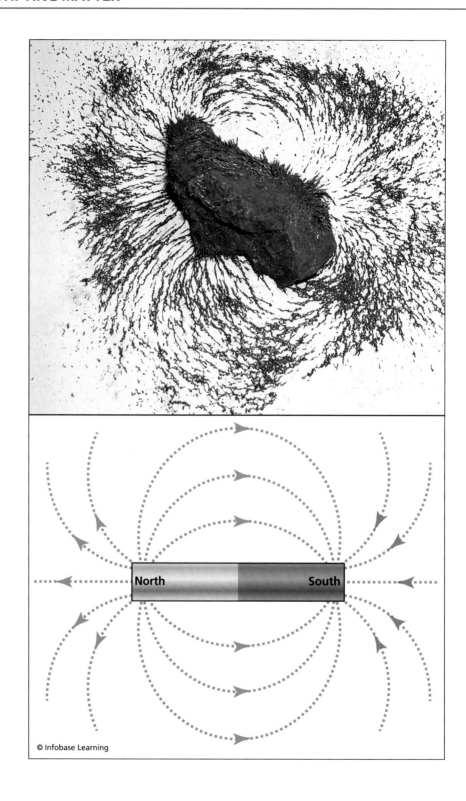

Gilbert's investigations of lodestone revealed many other important properties of natural magnets. His research led him to the very important conclusion that Earth is a giant magnet, thereby explaining the mysterious operation of the mariner's compass. He noted that every magnet has two poles, now arbitrarily called a north (or source) pole and a south (or sink) pole. As a result of his lodestone experiments, Gilbert discovered that like magnetic poles repel each other, while unlike magnetic poles attract each other. His experiments showed that north-south poles attract, while north-north and south-south poles repel.

The next major breakthrough in the physics of magnetism occurred quite serendipitously. At the University of Copenhagen, the Danish physicist Hans Christian Ørsted (1777–1851) was investigating what happens when an electric current flows through a wire. On April 21, 1820, he made preparations for laboratory demonstrations about electricity and about magnetism. He used the newly invented battery (Voltaic pile) as his source of electricity. As Ørsted moved his equipment around, he noticed that a current-carrying wire caused an unexpected deflection in the needle of a nearby compass. By chance, that particular wire was carrying a current because he was also planning to discuss how the flow electricity from a Voltaic pile could cause heating in a wire. The deflection of the nearby compass fortuitously caught his attention, and the event changed physics.

Over the next several months, Ørsted repeated the experiment and watched the compass needle deflect whenever current flowed through a nearby wire. He could not explain how the flow of electricity from a battery could cause this magnetic effect, but as a scientist, he felt obliged to report his unusual observations to Europe's scientific community. Ørsted's discovery had clearly linked electricity and magnetism for the very first time. The discovery immediately encouraged other scientists, such as the French physicist André-Marie Ampère (1775–1836), to launch their own comprehensive investigations of electromagnetism.

After Gilbert published *De Magnete*, other researchers turned their attention to magnetism. Some were fascinated by the fact that if they cut a

---

(opposite page) The top of this composite image shows a natural magnet, called lodestone, with a collection of iron filings that align along the special rock's magnetic field lines. The bottom drawing depicts an idealized bar magnet with its field lines extending from the magnet's north to south poles. *(Author [upper image]; Department of Energy [lower image])*

bar magnet in half, each half suddenly developed its own magnetic north and south poles. With this discovery, a more tantalizing technical question arose: Could scientists isolate a magnetic monopole—that is, a magnet with just a north pole or a south pole? Through the end of the 20th century, scientists searched for evidence of a magnetic monopole but failed. The contemporary consensus within today's scientific community is that nature does not support isolated magnetic poles. The detailed answer as to why this appears so lies in quantum mechanics and the characteristics of the magnetic fields associated with spinning electrons—specifically, the behavior of a material's magnetic dipole moments.

Each of an atom's electrons has an orbital magnetic dipole moment and a spin magnetic dipole moment. A material exhibits magnetic properties when the combination of each of these individual dipole moments produces a net magnetic field. Although a detailed, quantum-mechanical description of the phenomena that give rise to magnetism is beyond the scope of this book, the brief discussion that follows should prove helpful in understanding the magnetic properties of materials.

Scientists identify three general types of magnetism: ferromagnetism, paramagnetism, and diamagnetism. Ferromagnetic materials, such as iron, nickel, and certain other elements (including their compounds and alloys), have the resultant magnetic dipole moments aligned and consequently exhibit a strong magnetic field. These substances are normally referred to as magnetic materials or ferromagnetic materials. Quantum physicists suggest that a ferromagnetic substance enjoys the cooperative alignment of the electron spins of many atoms.

Materials containing rare earth elements, transition elements, and actinoid (formerly actinide) elements exhibit a property called paramagnetism. The magnetic dipole moments within such materials are randomly oriented. As a result, there is no net alignment that combines to create a significant magnetic field. In the presence of a strong external magnetic field, the individual magnetic dipole moments in a paramagnetic material tend to align and become weakly attracted by the external magnetic field. Quantum physicists suggest that this attraction is generally the result of unpaired electrons. The magnetic field of a paramagnetic substance, while observable, remains weaker than the magnetic field exhibited by ferromagnetic materials.

Finally, most common materials exhibit the property of diamagnetism. The atoms of diamagnetic substances produce only weak magnetic dipole moments. Even when placed in a strong external magnetic field, diamagnetic material exhibits only a feeble net magnetic field (if any). The diamagnetic substance is not attracted by an external magnetic field, or else the substance may be slightly repelled by this field. Quantum physicists suggest that diamagnetic substances have only paired electrons.

DISCOVERING ELECTRICITY: FROM ANCIENT GREECE TO COLONIAL AMERICA

The study of static electricity (electrostatics) extends back to the natural philosophers of ancient Greece. About 590 B.C.E., Thales of Miletus noted that amber, when rubbed, attracted small objects (such as pieces of grain or straw) to itself. The ancient Greeks called amber *electron*. Often used as jewelry because of its beauty, amber is typically yellow to orange in color and translucent. Amber is not a mineral but rather fossil resin (tree sap). Despite amber's mysterious property, any comprehensive investigation of static electricity languished for centuries.

At the start of the 17th century, Gilbert's *De Magnete* not only discussed magnetism but also stimulated renewed scientific interest in the phenomenon of electricity. Some scientists conducted experiments that generated static electricity by rubbing amber and other objects with cloth; other researchers created machines that could generate larger quantities of static electricity by friction. One such inquisitive individual was Otto von Guericke, who made an electrostatic generator in 1660. Von Guericke's *elektrisiermachine* used a rotating ball of sulfur and a handheld piece of cloth. More adventuresome experimenters would hold a bare hand against the rotating ball of sulfur. The machine functioned reasonably well. It accumulated an electric charge, generated sparks, and even gave an overly curious researcher a mild electric shock.

The French chemist and botanist Charles du Fay (1698–1739) wrote a paper in 1733 in which he discussed the existence of two types of electricity. He called one type *vitreous electricity* because it was the type of static electricity produced by rubbing glass with wool. (*Vitreous* is the Latin word for glass.) He called the other type *resinous electricity* because it was the type of static electricity produced by rubbing amber (fossilized resin). In a variety of experiments, du Fay was able to distinguish these two types of electricity from each other because an object that contained vitreous electricity would attract an object that contained resinous electricity but repel another object charged with vitreous electricity. Du Fay's work gave rise to the two-fluid theory of electricity, which was quite popular in the early to mid-18th century.

Scientists regard the Leyden jar (sometimes spelled Leiden) as the technical ancestor to the modern electric capacitor. In the mid-18th century, two individuals independently developed very similar devices to store the electric charge from friction generators. The first individual was Ewald Georg von Kleist (1700–48), a German cleric and scientist. He developed his device for storing electric charge in 1745 but failed to communicate his efforts to the scientific community in Europe. The Dutch scientist Pieter van Musschenbroek (1692–1761) independently developed a similar charge storage device at about the same time. Affiliated with the University of Leiden, Van Musschenbroek published details of his invention in January 1746 and then used it to conduct numerous experiments in electrostatics. This charge-storing device soon became widely known within the scientific community as the Leyden jar.

Lightning streaks across the night sky in the left portion of this composite illustration. The drawing on the right is a popular (though not technically accurate) picture of Benjamin Franklin's kite flying experiment of June 15, 1752. Franklin was either extremely lucky or, more likely, only collected some electric charge from an approaching storm cloud. A lightning strike on the kite would have killed him. Note: This is an extremely dangerous activity—do not attempt! (*National Park Service [left]; Department of Energy [right]*)

In its most basic form, a Leyden jar consists of a glass jar that has its inner and outer surfaces coated with electrically conducting metal foil. To prevent electric arcing between the foils when charged, the conducting materials stop just short of the jar's top. A well insulated electrode (such as a metal rod and chain arrangement) passed through the mouth of the jar and extended down into it, enabling electrical contact with the inner foil. Scientists charged the Leyden jar by touching this central electrode to an electrostatic generator. As researchers attempted to store more and more charge in support of their investigations of electricity, many different designs appeared in the late 18th century.

The American statesman, printer, and patriot Benjamin Franklin (1706–90) was also an excellent scientist. His technical achievements were well recognized in Europe, especially in France. Franklin began his investigation of electricity in 1746. During experiments with a simple capacitorlike array of charged glass plates in 1748, he coined the term *battery*. Franklin used that term because his configuration of charge-storing devices resembled an artillery battery.

Franklin dismissed du Fay's two-fluid theory of electricity and replaced it with his own one-fluid theory. He was convinced that all the phenomena associated with the transfer of electric charge from one object to another could easily be explained by the flow of just a single electric fluid. His experiments suggested that objects either had an excess of electricity or too little electricity. To support his one-fluid theory, Franklin introduced the terms *negative* (too little) and *positive* (too much) to describe these conditions of electric charge. Franklin also coined many other electricity-related terms that are now in use. Some of these terms are *electric shock, electrician, conductor, condenser, charge, discharge, uncharged, plus charge,* and *minus charge.*

When asked to describe Franklin's role as a scientist, most people mention his historic kite-flying experiment, by which he demonstrated that lightning was electrical in nature. Although Franklin's discovery was a major accomplishment, it did not take place as generally depicted in textbooks and paintings. Flying a kite into a thunderstorm is an extremely dangerous activity. Had lightning actually struck Franklin's kite during his famous experiment on June 15, 1752, he would most likely have been electrocuted. What Franklin did during the experiment was to fly his kite into a gathering storm cloud—*before* any lightning discharges began. He detected the buildup of an electric charge within the cloud, mentally associated charge buildup with lightning, quickly ceased the experiment, and headed for safety. Franklin was definitely aware of the serious danger posed by lightning—he invented the lightning rod to protect structures.

A contemporary researcher named Georg Wilhelm Richmann (1711–53) was not so lucky and became the first person known to be electrocuted by lightning during a scientific experiment. While in Saint Petersburg, Russia, the German scientist attempted to measure the performance of an insulating rod during a lightning storm. According to historic reports, as Richmann conducted this dangerous experiment on August 6, 1753, lightning traveled down the equipment and entered his head. He died instantly.

By the end of the 18th century, scientists had determined the following important facts about electricity: (1) like electric charges repel, while unlike (positive and negative in Franklin's nomenclature) charges attract; (2) positive and negative charges appear to have equal strength; and (3) lightning is an electrical phenomenon. Nevertheless, no one had even a hint that the true nature of electricity involved the flow of electrons.

COUNT ALESSANDRO VOLTA

It may be difficult to imagine that all modern applications of electromagnetism descended from the availability of a dependable supply of electricity, the first of which appeared about 210 years ago. This section discusses how the Italian physicist Count Alessandro Volta (1745–1827) performed a series of key experiments in 1800 that led to the development of the first electric battery. Early batteries enabled the 19th-century research activities that led to the electricity-based Second Industrial Revolution.

Volta was born on February 18, 1745, in Como, Italy. Like many other 18th-century scientists, Volta became fascinated with the subject of electricity and decided to focus his research activities on a detailed investigation of this mysterious natural phenomenon. In 1779, Volta received an appointment to become a professor of physics at the University of Pavia. He accepted this appointment and remained in this position for the next 25 years.

Volta had a friend and professional acquaintance named Luigi Galvani (1737–98) who was a physician (anatomist) in Bologna, Italy. In 1780, Galvani discovered that when a dissected frog's leg touched two dissimilar metals (such as iron and brass or copper and zinc) at the same time, the leg twitched and contracted. Based on the observed muscular contractions,

Galvani postulated that the flow of electricity in the frog's leg represented some type of *animal electricity,* a term he devised to suggest that electricity was the animating agent in living muscle and tissue. Galvani may have been influenced by Franklin's research, which associated lightning (a natural phenomenon) to electricity. Electricity was a frontier science in the late 18th century, so Galvani, as a scientist with a strong inclination toward anatomy, wanted to be the first investigator to successfully connect the animation of living matter with this exciting "new" phenomenon.

ELECTRICITY AND LIVING THINGS

During the 1790s, the Italian physiologist Luigi Galvani actively explored a phenomenon he called animal electricity—a postulated life-giving electric force presumed capable of animating inanimate matter. Although Alessandro Volta eventually proved that animal electricity did not exist, Galvani's research did anticipate the discovery of nervous impulses (which travel through the living body) and the existence of tiny electric currents in the brain (which are now noninvasively measured to support medical research and diagnosis).

Electrotherapy was practiced by medical doctors and scientists from the late 1790s through the mid-20th century. Physicians used batteries or small electric generators in their attempts to apply electric currents to cure various diseases and mental conditions. Since few reliable cures were actually achieved, electrotherapy has all but vanished from the landscape of modern medicine.

In the early 1800s, scientists and physicians were intrigued by the elusive boundary between life and death. They performed many experiments attempting to resuscitate drowning victims with procedures that often included the use of electricity. Historians suggest that these activities and Galvani's pursuit of animal electricity may have provided the intellectual stimulus for Mary Shelley's (1797–1851) iconic horror story *Frankenstein,* which was first published in 1818.

Contemporary medical practitioners employ a variety of sensitive instruments to measure the body's electrical impulses. With such electrical signals, physicians can routinely monitor the heart, brain, and other important organs. In addition, pacemakers help correct irregular heartbeats, and defibrillators can help restore the heart's natural rhythm after a person has experienced cardiac arrest or a dangerous arrhythmia.

Galvani knew that Volta was also performing experiments with electricity, so he asked Volta to help validate his experiments and the hypothesis about animal electricity. This request eventually ended the amicable relationship between the two Italian scientists and made all their subsequent interactions adversarial.

In about 1794, Volta began to explore the question of whether the electric current in the twitching frog legs was a phenomenon associated with biological tissue (as Galvani postulated) or actually the result of contact between two dissimilar metals. Ever the careful physicist, he tested two dissimilar metals alone, without a frog's leg or other type of living tissue. He observed that an electric current appeared and continued to flow. The frog's leg had nothing to do with the current flow. Volta's conclusions dealt a mortal blow to Galvani's theory of animal electricity.

Galvani did not accept Volta's conclusions, and the two Italian scientists engaged in a bitter controversy that soon involved other famous scientists from across Europe. The French physicist Charles Augustin de Coulomb (1736–1806) supported Volta's work and conclusions. As additional experimental evidence began to weigh heavily in Volta's favor, Galvani died a broken and bitter man on December 4, 1798. To the end, Galvani clung to his belief that electricity was inseparably linked to biology. Like many other investigators of the day, Galvani regarded electricity as a natural agent that promoted vitality.

The professional disagreement with Galvani encouraged Volta to perform additional experiments. In 1800, Volta developed the voltaic pile—the first chemical battery. From a variety of experiments, Volta determined that in order to produce a steady flow of electricity, he needed to use silver and zinc as the most efficient pair of dissimilar metals. He initially made individual cells by placing a strip of zinc and silver into a cup of brine. He then connected several cells to increase the voltage. Finally, he created the first voltaic pile (chemical battery) by alternately stacking discs of silver, zinc, and brine-soaked heavy paper—quite literally in a pile. Soon, scientists all over Europe copied and improved upon Volta's invention. The early batteries provided a steady, dependable flow of electricity (direct current) for numerous pioneering experiments. Volta's battery initiated a revolution in the scientific investigation of electromagnetism.

In 1810, Emperor Napoleon of France acknowledged Volta's great accomplishment and made him a count. Volta died in Como, Italy, on March 5, 1827. Modern batteries are distant technical descendants of Volta's electric pile. In their numerous shapes and forms, modern batteries continue to electrify civilization. Scientists named the SI unit of electric potential difference and electromotive forces the volt (V) to honor Volta's great contributions to science.

QUANTIFYING THE ELECTRIC NATURE OF MATTER

This section discusses the contributions of three key individuals (Coulomb, Ampère, and Ohm) who helped the scientific community better understand electricity. Their scientific contributions paved the way for the electricity-based Second Industrial Revolution.

The French military engineer and scientist Charles Augustin de Coulomb performed basic experiments in mechanics and electrostatics in the late 18th century. In particular, he determined that the electrostatic force, which one point charge applies to another, depends directly on the amount of each charge and inversely on the square of their distance of separation. In his honor, the SI unit of electric charge is called the coulomb (C). Scientists define one coulomb as the quantity of electric charge transported in one second by a current of one ampere (A).

Building upon Coulomb's important work, experiments by physicists in the 20th century revealed that the magnitude of the elementary charge (e) on the proton exactly equals the magnitude of the charge on the electron. By convention, scientists say the proton carries a charge of $+e$ and the electron carries a charge of $-e$. Scientists have experimentally determined the value of e to be 1.602×10^{-19} C.

Science historians regard André-Marie Ampère as one of the main discoverers of electromagnetism. The gifted French scientist's defining work in this field began in 1820 and involved insightful experiments that led to the development of a physical principle called Ampère's law for static magnetic fields. His pioneering work in electromagnetism served as the foundation of the subsequent work of the British experimenter Michael Faraday and the American physicist Joseph Henry (1797–1878). The science of electromagnetism helped bring about the world-changing revolution in electric power applications and information technology that characterized the late 19th century. In his honor, the SI unit of electric current is called the ampere (A), or amp for short.

The German physicist George Simon Ohm (1787–1854) published the results of his experiments with electricity in 1827. His research suggested the existence of a fundamental relationship between voltage, current, and resistance. Despite the fact that his scientific colleagues dismissed these important findings, Ohm's pioneering work defined fundamental physical relationships that supported the beginning of electrical circuit analysis. Ohm proposed that the electrical resistance (R) in a material could be defined as the ratio of the voltage (V) applied across the material to the electric current (I) flowing through the material, or R = V/I. Today, physicists and engineers refer to this important relationship as Ohm's law. In recognition of his contribution to the scientific understanding of electricity, scientists call the SI unit of resistance the ohm (symbol Ω). One ohm of electric resistance is defined as one volt per ampere (1 Ω = 1 V/A).

MICHAEL FARADAY'S REVOLUTION

Even though he lacked formal education and possessed only limited mathematical skills, Michael Faraday became one of the world's greatest experimental scientists. Of particular interest is the fact that the British scientist made significant contributions to the field of electromagnetism. In 1831, he observed and carefully investigated the principle of electromagnetic induction, an important physical principle that governs the operation of modern electric generators and motors. In his honor, scientists named the SI unit of capacitance the farad (F). One farad is defined as the capacitance of a capacitor whose plates have a potential difference of one volt when charged by a quantity of electricity equal to one coulomb. Since the farad is too large a unit for typical applications, submultiples— such as the microfarad (μF), the nanofarad (nF), and the picofarad (pF)— are frequently encountered in electrical engineering.

Faraday's pioneering work (ca. 1831) in electromagnetism formed the basis of modern electric power generation. He discovered that whenever there is a change in the flux through a loop of wire, an electromotive force *(emf)* is induced in the loop. Scientists now refer to this important discovery as Faraday's law of electromagnetic induction. It is the physical principle upon which the operation of an electric generator (dynamo) depends. Faraday's discovery, refined by electrical engineers into practical generators, made large quantities of electricity suddenly available for research and industrial applications. Scientists were no longer restricted to electricity supplied by chemical batteries.

Independently of Faraday, the American physicist Joseph Henry had made a similar discovery about a year earlier, but teaching duties prevented Henry from publishing his results, so credit for this discovery goes to Faraday, who actually not only published his results first (*Experimental Researches in Electricity, first series* [1831]), but also performed more detailed experimental investigations of the important phenomenon. Henry did publish a seminal paper in 1831, however, that described the electric motor (essentially a reverse dynamo) and its potential applications. Science historians regard the work of both Faraday and Henry during this period as the beginning of the electrified world. Clever engineers and inventors would apply Faraday's law of electromagnetic induction to create large electric generators that supply enormous quantities of electricity. Other engineers would invent ways of using direct current (DC) and alternating current (AC) electricity to power the wide variety of practical and efficient electric motors that became the "motive force" of modern civilization.

Faraday was ingenious in his design and construction of experiments. However, he lacked a solid mathematics education, so translating the true significance of some of his experimental results into robust physical theory depended upon his affiliation with the Scottish theoretical physicist James Clerk Maxwell. Maxwell, a genius in his own right, competently translated the significance of Faraday's ingenious experiments into the mathematical language of physics. Their cordial working relationship resulted in a solid experimental and mathematical basis for classical electromagnetic theory.

MAXWELL'S ELECTROMAGNETIC THEORY

The Scottish mathematician James Clerk Maxwell is regarded as one of the world's greatest theoretical physicists. Born in Edinburgh in 1831, he was soon recognized as a child prodigy. Throughout his brilliant career he made many contributions to science, including papers that linked the behavior of individual atoms and molecules to the bulk behavior of matter—an area of classical physics known as kinetic theory.

His seminal work, *Treatise on Electricity and Magnetism* (published in 1873), provided a detailed mathematical treatment of electromagnetism. Maxwell suggested that oscillating electric charges could generate propagating electromagnetic waves that had the speed of light as their propagation speed. Maxwell's equations provided scientists a comprehensive insight into the physical nature of electromagnetism. His work provided a detailed model of visible light as a form of electromagnetic radiation. Maxwell also suggested that other forms of electromagnetic radiation might exist beyond infrared and ultraviolet wavelengths. Scientists such as Rudolph Hertz (1857–94) and Wilhelm Conrad Roentgen (1845–1923) would soon provide experimental evidence that proved Maxwell correct.

In 1888, the German physicist Heinrich Hertz oscillated the flow of current between two metal balls separated by an air gap. He observed that each time the electric potential reached a peak in one direction (or the other), a spark would jump across the gap. Hertz applied Maxwell's electromagnetic theory to the situation and determined that the oscillating spark should generate an electromagnetic wave that traveled at the speed of light. He also used a simple loop of wire with a small air gap at one end to detect the presence of electromagnetic waves produced by his oscillating spark circuit. With this pioneering experiment, Hertz produced and detected radio waves for the first time. He also demonstrated that (as predicted by Maxwell) radio waves propagated at the speed of light. Hertz showed that radio waves were simply another form of electromagnetic radiation, similar to visible light and infrared radiation save for their longer wavelengths and lower frequencies.

Hertz's research formed the basis of the wireless communications industry. Scientists named the SI unit of frequency the hertz (Hz) in his honor. One hertz is equal to one cycle per second.

On November 8, 1895, the German physicist Wilhelm Conrad Roentgen discovered X-rays, another form of electromagnetic radiation. (See chapter 7.) With shorter wavelengths and higher frequencies, X-rays lie beyond ultraviolet radiation in the electromagnetic spectrum.

Maxwell's brilliant theoretical interpretation of electromagnetism became the intellectual bridge between the classical physics of the 19th century and modern physics of the 20th century. When Maxwell died on November 5, 1879, in Cambridge, England, he was only 48 years old. Many scientists regard his contributions to physics as comparable to those of Newton and Einstein.

SIR JOSEPH JOHN (J. J.) THOMSON DISCOVERS THE ELECTRON

At the end of the 19th century, the British physicist Sir Joseph John (J. J.) Thomson made a monumental discovery. Using Crookes tubes (cold cath-

THE ELECTROMAGNETIC SPECTRUM

When sunlight passes through a prism, it throws out a rainbowlike array of colors onto a surface. Scientists call this display of colors the *visible spectrum.* The visible spectrum consists of the narrow band of electromagnetic (EM) radiation to which the human eye is sensitive.

However, the EM spectrum consists of the entire range of EM radiation, from the shortest wavelength gamma rays to the longest wavelength radio waves and everything in between. EM radiation travels at the speed of light and represents the basic mechanism for energy transfer through the vacuum of outer space.

One of the most interesting discoveries of 20th-century physics is the dual nature of EM radiation. Under some conditions, EM radiation behaves like a

(continues)

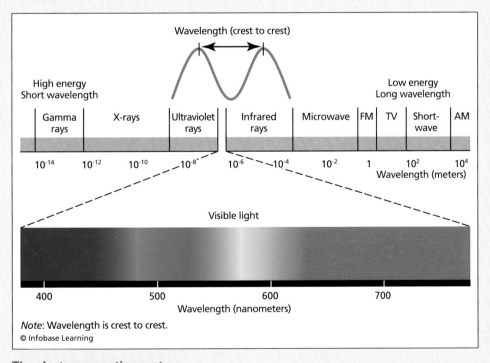

The electromagnetic spectrum

(continued)

wave, while under other conditions, it behaves like a stream of particles called *photons.* The tiny amount of energy carried by a photon is called a quantum of energy.

The shorter the wavelength, the more energy is carried by a particular form of EM radiation. All things in the universe emit, reflect, and absorb EM radiation in their own distinctive ways. How an object does this provides scientists a spectral signature that can be detected by remote sensing instruments. Scientists often collect spectral signatures in order to obtain information about an object's composition, density, surface temperature, and/or motion.

ode tubes), he performed experiments that demonstrated the existence of the first known subatomic particle, the electron.

Thomson's discovery implied that the previously postulated indivisible, solid atom was actually divisible and contained smaller parts. When he announced his finding during a lecture at the Royal Institution in 1897, his audience was initially reluctant to accept the idea that objects smaller than an atom could exist. Even so, Thomson was correct, and he received the 1906 Nobel Prize in physics for his landmark discovery. He was knighted in 1908.

Thomson originally referred to the tiny subatomic particles he discovered as *corpuscles.* Several years earlier (in 1891), the Irish physicist George Johnstone Stoney (1826–1911) suggested the name *electron* for the elementary charge of electricity. In time, Stoney's term became the universal name by which Thomson's tiny negatively charged corpuscles became known. The existence of the electron showed that (some) subatomic particles carried a fundamental (quantized) amount of electric charge.

In 1898, Thomson became the first scientist to put forward a technical concept concerning the interior structure of the atom. He suggested the atom was actually a distributed positively charged mass with an appropriate number of tiny electrons embedded in it, much like raisins in a plum pudding. Because there were still so many unanswered questions, not many physicists rushed to embrace Thomson's model of the atom, commonly referred to as the plum pudding model, but the Thomson atom, for all its limitations, started Ernest Rutherford (1871–1937) and other atomic scientists thinking about the structure within the atom. (See chapter 8.)

Today, scientists regard the electron (e) as a stable elementary particle with a fundamental unit of negative charge (-1.602×10^{-19} C) and a rest mass of just 2.009×10^{-30} lbm (9.109×10^{-31} kg)—a mass approximately 1/1,837 that of a proton. Scientists now recognize that electrons surround the positively charged nucleus of an atom and determine a substance's chemical properties and behavior. The manipulation and flow of electrons support today's information-dependent global civilization.

CHAPTER 6

Periodic Table of the Elements

This chapter explores the origin, content, and applications of the periodic table of the chemical elements. As described in this chapter and in the Appendix, the modern version of Mendeleev's periodic table is greatly expanded in content and has the elements arranged by each element's atomic number.

NEW CHEMICAL ELEMENTS KEEP SHOWING UP

As alchemy transitioned into the science of chemistry during the 18th century, scientists identified a large number of new elements. Chemists began debating about how these new chemical substances were related. The fact that the new materials were appearing at a rather alarming rate further complicated the situation. To highlight the growing quandary, the following is an alphabetical listing of the 19 new elements that were identified during the 18th century:

barium (Ba)
beryllium (Be)
chlorine (Cl)
chromium (Cr)
cobalt (Co)

hydrogen (H)
magnesium (Mg)
manganese (Mn)
molybdenum (Mo)
nickel (Ni)

nitrogen (N)
oxygen (O)
platinum (Pt)
strontium (Sr)
tellurium (Te)

tungsten (W)
uranium (U)
yttrium (Y),
zirconium (Zr)

In the first six decades of the 19th century, scientists discovered 28 new chemical elements. These elements are listed in the following table along with the year of discovery:

vanadium (V)	(1801)	lithium (Li)	(1817)
niobium (Nb)	(1801)	cadmium (Cd)	(1817)
tantalum (Ta)	(1802)	selenium (SE)	(1817)
palladium (Pd)	(1802)	silicon (Si)	(1824)
cerium (Ce)	(1803)	aluminum (Al)	(1825)
osmium (Os)	(1803)	bromine (Br)	(1825)
iridium (Ir)	(1803)	thorium (Th)	(1829)
rhodium (Rh)	(1804)	lanthanum (La)	(1838)
potassium (K)	(1807)	erbium (Er)	(1842)
ruthenium (Ru)	(1807)	cesium (Cs)	(1860)
sodium (Na)	(1807)	rubidium (Rb)	(1861)
calcium (Ca)	(1808)	thallium (Tl)	(1861)
boron (B)	(1808)	indium (In)	(1863)
iodine (I)	(1811)	helium (He)	(1868)

How scientists discovered each of these new elements forms a collection of interesting stories that unfortunately lies well beyond the size limitations of this chapter. However, the scientific efforts of one individual, the Swedish chemist Jöns Jacob Berzelius, will be discussed here. His contributions provide good insight into the heightened intellectual activity characteristic of the pre–periodic table era in chemistry.

Berzelius was born in 1779 in Väversunda, Sweden. Having lost both his parents by the age of nine, he spent the early part of his life in poverty and distress. Overcoming these impediments, he studied medicine at Uppsala University and received his degree in 1802. One of his instructors was a chemistry professor named Anders Gustaf Ekeberg (1767–1813). Ekeberg was a Swedish chemist who discovered the element tantalum (Ta). In 1803, while working with Swedish chemist Wilhelm Hisinger (1766–1852), Berzelius discovered the element cerium (Ce). Through his association with Hisinger, Berzelius acquired a laboratory and soon demonstrated that all salts can be decomposed by electricity.

Berzelius expanded his activities in chemistry in 1807 by writing an excellent textbook for his students. Foreign travel over the next two decades brought him into personal contact with some of Europe's leading chemists. In 1815, he received an appointment to serve as a professor of chemistry at the Royal Caroline Medico-Chirurgical Institute (Karolinska Institute) in Stockholm and retained that position until he retired in 1832.

In addition to the codiscovery of cerium, subsequent research allowed Berzelius to discover the following elements: selenium (Se) in 1817, silicon (Si) in 1824, and thorium (Th) in 1829. He was also the first scientist to isolate silicon, titanium, and zirconium and engaged in detailed studies of other rare metals.

His greatest contribution to chemistry took place between 1807 and 1817, when he analyzed some 2,000 compounds and began to puzzle over the atomic nature of matter. His work led him to speculate whether the chemical elements were always present in compounds in fixed proportions. This issue was subject to great debate, stimulated for the most part by the work of Proust and Dalton.

The French analytical chemist Joseph Louis Proust (1754–1826) had stated in 1799 that regardless of the method employed the proportions of the constituents in any chemical compound are always the same. His meticulous analysis of copper carbonate, whether prepared in the laboratory or by natural processes, showed Proust that this chemical compound always contained (by mass): 57.5 percent copper, 5.4 percent carbon, 0.9 percent hydrogen, and 36.2 percent oxygen. Scientists called this important scientific observation Proust's law of definite proportions.

In 1803, Dalton suggested that each chemical element was composed of a particular type of atom. His interest in the behavior of gases allowed him to further investigate the atomic concept of matter. Dalton proposed that the relative masses, or weights, of different atoms could be determined. In establishing his relative scale, he assigned the atom of hydrogen a mass of unity. Dalton summarized this work in the 1808 publication *New Systems of Chemical Philosophy*. His work revived the concept of atomism and encouraged other scientists to investigate the hypothesis that all matter is made up of combinations of atoms.

Berzelius combined Proust's notion of constant proportions with Dalton's revival of atomic theory and went on to develop a table of atomic weights, using the element oxygen as his reference. (Historically, chemists defined *atomic weight* as the mass of an atom [in atomic mass units] relative to other atoms.) Berzelius published his first table of atomic weights in 1818. The table contained 45 elements. He later revised the table in 1828. The revised table had 54 elements. Although the values for atomic weights expressed in Berzelius's table agreed reasonably well with the modern values of atomic masses, there were some notable discrepancies.

The main reason lies in the fact that most elements have just one major isotope that contributes to the overall atomic mass, but there are a few elements, such as chlorine (with an atomic mass of 35.45), that have several stable isotopes that contribute to the element's overall atomic mass. These circumstances can drive the average mass away from an anticipated near-integer value. The two major stable isotopes of chlorine are Cl-35, with a natural abundance of 75.7 percent, and Cl-37, with a natural abundance of 24.2 percent. In Berzelius's time, scientists had not yet developed the concept of an isotope, so no one could explain discrepancies in atomic weight values relative to anticipated integer values.

Furthermore, Berzelius did not appreciate the true significance of Avogadro's hypothesis. Like other scientists of the time, he probably was somewhat confused as to the difference between atoms and elemental molecules. For example, a hydrogen *molecule* (H_2) consists of two atoms of hydrogen. In the end, most of the discrepancies that appeared in Berzelius's table of atomic weights were reasonably resolved in 1860 by the Italian chemist Stanislao Cannizzaro (1826–1912), who recognized the difference between atomic weight and molecular weight.

While developing his table of atomic weights, Berzelius devised the modern system of symbols for the chemical elements. Thinking in terms of atoms, Berzelius wrote chemical formulas using the first letter (or two letters as necessary to avoid confusion) of the name of the chemical elements in a given molecule. Scientists still use this system. The only subtle difference is the fact that Berzelius used superscripts to indicate the relative number of atoms of an element in a particular molecule, while modern scientists use subscripts. Although Dalton opposed this type of notation, other chemists adopted it, and Berzelius's system went on to became the international symbolic language of chemistry.

Berzelius was a very skilled experimenter, and many of his laboratory devices are still part of the equipment complement in a standard chemical laboratory. His equipment included rubber tubing, wash bottles, and filter paper. He also receives credit for coining such terms as *allotrope, isomer,* and *polymer.* An allotrope represents one of two or more distinct forms of a pure chemical element in the same physical state. Isomers are compounds that have the same chemical formula but possess different physical arrangements of their atoms. A polymer is a very large molecule that consists of a number of smaller molecules linked together repeatedly by covalent bonds.

At his retirement in 1832, Berzelius was regarded as the most famous chemist in Europe. Science historians rank him alongside Lavoisier and Dalton as a founding member of modern chemistry. He died in Stockholm in 1848.

SEARCHING FOR A PERIODIC LAW IN CHEMISTRY

During the first half of the 19th century, scientists made numerous attempts to arrange the growing number of chemical elements into some logical order. Dalton had proposed a table of relative atomic masses using hydrogen (with an atomic mass of one) as his reference. Berzelius improved Dalton's efforts by proposing oxygen as the reference in his own table of atomic weights.

Starting around 1816, the German chemist Johann Wolfgang Döbereiner (1780–1849) tried to organize the known elements into simpler, more understandable subgroups. By 1829, he observed that elements with similar physical and chemical properties appeared to fall into triads, or groups of three. One of Döbereiner's triads consisted of chlorine (Cl), bromine (Br), and iodine (I); another group of three involved calcium (Ca), strontium (Sr), and barium (Ba); another chemical triad contained sulfur (S), selenium (Se), and tellurium (Te); and still another group of three involved lithium (Li), sodium (Na), and potassium (K). In each case, Döbereiner noticed that the atomic weight of the middle element in a triad was approximately the average atomic weight of the other two elements in the group. The middle element in each triad also exhibited about half the density and other averaged properties of the other two members. Döbereiner's triad model quickly became obsolete when several newly discovered elements clearly did not fit into his proposed model.

In 1862, the French geologist Alexandre-Emile Béguyer de Chancourtois (1820–86) also arranged the known chemical elements in order of their atomic weights. His approach to classifying the chemical elements incorporated the latest values of atomic weights recently presented by Cannizzaro at the 1860 scientific conference in Karlsruhe, Germany. Béguyer de Chancourtois used an innovative spiral graph that he wrapped around a cylinder to display the elements in order of their atomic weights. Rather cleverly, he plotted the atomic weights on the cylinder such that one complete turn (or revolution) on his spiral graph corresponded to an atomic weight increase of 16. The element tellurium (Te), with an atomic weight of 127.6, appeared in the middle of the spiral graph.

Once he had plotted all the known elements on this unusual spiral graph, he observed that elements with similar properties appeared to align vertically, suggesting some type of periodicity. Possibly because he was a geologist or possibly because he did not provide an image of the spiral graph in his 1862 paper to the French Academy of Sciences (which announced his findings), chemists around Europe ignored his results. As a result of this oversight, the honor of developing a well-recognized periodic law for the chemical elements awaited the efforts of a Russian chemist named Dmitri Mendeleev.

Starting in 1863, the British industrial chemist John Alexander Newlands (1837–98) built upon the work of Berzelius and Döbereiner in his own effort to bring order to the number of chemical elements and their different chemical and physical properties. By 1865, Newlands had constructed a table in which he arranged the chemical elements in octaves (groups of eight) in order of ascending values of atomic weight. He noticed that every eighth element in his table had similar chemical properties. This discovery encouraged Newlands to suggest his law of octaves. Unfortunately, the noble gases had not yet been discovered, and this convenient grouping of eight began to encounter serious problems beyond the element calcium (Ca). Most chemists quickly rejected the octave model, since it could not provide a credible periodic relationship for all the known elements.

In 1864, the German chemist Julius Lothar von Meyer (1830–95) wrote a textbook on chemistry. As part of this endeavor, he constructed a table of 28 chemical elements by arranging these elements into six families according to their atomic weights and chemical reactivities (or combining powers), which scientists now term *valence*. Sodium (Na) has a valence of one and combines with chlorine (Cl) to form sodium chloride (NaCl), better known as salt. Magnesium (Mg) has a valence of two and combines with two chlorine atoms to form magnesium chloride ($MgCl_2$). The six chemical families in von Meyer's table displayed similar chemical and physical properties. Independent of Mendeleev, von Meyer continued to work on his valence-based periodic table of elements and published a refined and expanded set of results in 1870, the year after Mendeleev introduced his own famous law and table of periodicity. In 1882, the British Royal Society presented both von Meyer and Mendeleev the prestigious Davy medal in recognition of their independent efforts that resulted in the development of the periodic table.

MENDELEEV'S PERIODIC LAW

The Russian chemist Dmitri Ivanovich Mendeleev was born in 1834 in Tobolsk, Siberia. In 1849, his widowed mother moved the family to Saint Petersburg, where he enrolled at the Main Pedogogical Institute. Following graduation from college in 1855, Mendeleev traveled to the Crimean Peninsula for health reasons. He returned to Saint Petersburg in 1857, having fully recuperated from his bout with tuberculosis. As a young chemist, he participated in the 1860 scientific conference held in Karlsruhe, Germany. The presentation by Cannizzaro regarding atomic weights greatly impressed Mendeleev, and he became interested in searching for relationships among the growing number of chemical elements.

In 1862, he married his first wife (Feozva Nikiticha Leshcheva) and the following year started teaching chemistry at the Saint Petersburg Technological Institute. He joined the faculty at the University of Saint Petersburg in 1866. To assist his students, he published a classic chemistry textbook, *The Principles of Chemistry,* in 1869. In this textbook, Mendeleev introduced the important principle he called the periodic law of chemical elements. This law stated that the properties of the elements are a periodic function of their atomic weights. Other chemists had tried in vain to bring some type of order to the growing collection of elements. Mendeleev's attempt was the most successful arrangement to date and the one soon embraced by chemists around the world. He presented his initial version of the periodic table in 1869 and a revised version in 1871.

Like others before him, Mendeleev began by arranging his table of elements with each row going (from left to right) primarily in order of increasing atomic weights. He also used an element's chemical reactivity (or valence) as a figure of merit to help define the population of each column (group). Examining the rise and fall of valence values as a function of atomic weight, he grasped the important fact that there was a certain periodicity in chemical properties. In order to better align his table in a few places, he positioned a slightly heavier element before a lighter one. This decision allowed him to maintain similar chemical properties in each vertical column (group). For example, in one case, he placed the element tellurium (Te), which has an atomic weight of 127.6, ahead of iodine (I), which has an atomic weight of 126.9. By so doing, tellurium appeared in the same column (group) as sulfur (S) and selenium (Se)—all three of which elements have similar chemical properties. Mendeleev also positioned iodine in the same column (group) as chlorine (Cl) and bromine (I), clearly another good match of valences and similar properties.

His other innovative decision was to leave gaps in the table. These gaps provided places for the discovery of anticipated new elements. An excellent chemist, he boldly predicted the properties that some of these missing elements should possess. He referred to certain missing elements as ekaboron, ekaaluminum, ekamanganese, and ekasilicon. (The prefix *eka* is an adjective meaning "one" from the historic Indo-Aryan language Sanskrit.) Mendeleev used the term *ekaaluminum* to identify a predicted but yet to be discovered element residing one place down from aluminum in his periodic table.

Mendeleev proposed that ekaboron would have an atomic weight of 44, ekaaluminum an atomic weight of 68, ekasilicon an atomic weight of 72, and ekamanganese an atomic weight of 100. Mendeleev's brilliant work was initially met with skepticism. However, as soon as several of his

This Polish stamp honors the Russian chemist Dmitri Ivanovich Mendeleev. *(Author)*

predicted ekaelements were discovered with properties closely approximating his projections, Mendeleev's periodic table rapidly became the Rosetta stone of chemistry. Scientists throughout the world recognized him as Russia's greatest chemist.

The French chemist Paul Emile Lecoq de Boisbaudran (1838–1912) discovered the element gallium (Ga) in 1875. Mendeleev predicted that ekaaluminum would have an atomic weight of 68 and a density of 374.6 lbm/ft^3 (6.0 g/cm^3). Gallium has an atomic weight of 69.72 and a density of 368.3 lbm/ft^3 (5.9 g/cm^3). Lecoq de Boisbaudran's discovery provided an enormous boost to Mendeleev's periodic table. Gallium lies just below aluminum on the modern periodic table.

The Swedish chemist Lars Fredrik Nilson (1840–99) discovered the rare earth element scandium (Sc) in 1879. Scandium has an atomic weight of 44.96. Mendeleev's ekaboron had a predicted atomic weight of 44. Similarly, the German chemist Clemens Winkler (1838–1904) discovered the element germanium (Ge) in 1886. Germanium has an atomic weight of 72.64 and a density of 334.0 lbm/ft^3 (5.35 g/cm^3). Mendeleev's ekasilicon had a predicted atomic weight of 72 and a predicted density of 343.4 lbm/ft^3 (5.5 g/cm^3). Winkler's discovery reinforced the validity of Mendeleev's work. In 1937, the Italian scientist Emilio Segrè (1905–89) discovered the element technetium (Tc), the most long-lived radioisotope of which is Tc-98. Technetium has an approximate atomic weight of 98. In 1871, Mendeleev postulated that ekamanganese would have an atomic weight of about 100.

While the outside world viewed Mendeleev as a brilliant chemist, within czarist Russia, his brilliant contributions to science were overshadowed by scandalous personal behavior and unacceptable political activism. After several years of open infidelity to his first wife, he finally divorced her and married Anna Ivanova Popova in 1882. He failed to gain admission to the Russian Academy of Sciences. Czarist Russia was extremely conservative, and government and religious officials had little tolerance for social activism or openly scandalous marital relationships. Although recognized all over Europe as one of the world's outstanding chemists, academic officials forced his resignation from Saint Petersburg University in 1890.

Despite this personal setback, Mendeleev remained innovative and vibrant. Three years later, he accepted a position as director of Russia's Bureau of Weights and Measures. Not only did he use this position to introduce the metric system into Russia but also endeared himself to every vodka-drinking citizen by formulating new state standards for the production of that alcoholic beverage. Under the Russian law passed in 1894, all vodka produced in Russia had to be 40 percent alcohol by volume. Mendeleev died in 1907 in Saint Petersburg.

MODERN PERIODIC TABLE OF ELEMENTS

The chemical properties of matter remained a mystery until woven into an elegant intellectual tapestry called the periodic table. This section briefly describes the content and significance of the periodic table, which appears in the appendix of this book.

The modern periodic table evolved from Mendeleev's pioneering work. However, several major scientific developments—such as the discovery of radioactivity, the discovery of the electron, the introduction of the nuclear atom model, and the emergence of quantum mechanics—were necessary before the periodic table could assume its current structure and detailed information content. Today, scientists continue to explore the nature of matter as they attempt to create new, super-heavy elements.

As of July 2010, scientists have observed and identified 118 elements. Of these, 94 occur naturally here on Earth. However, six of these 94 elements exist naturally only in trace quantities. These fleeting elements are technetium (Tc) (atomic number 43), promethium (Pm) (61), astatine (At) (85), francium (Fr) (87), neptunium (Np) (93), and plutonium (Pu) (94). The advent of nuclear accelerators and reactors has made synthetic production of these scarce elements possible. Some (such as neptunium and plutonium) are now available in significant quantities. The nuclear age has also made available countless relatively short-lived radioisotopes of all the elements.

Chemists and physicists now correlate the properties of the elements portrayed in the periodic table by means of their electron configurations. Since in a neutral atom the number of electrons equals the number of protons, scientists find it convenient to arrange the elements in the periodic table in order of their increasing atomic number. The modern periodic table has seven horizontal rows (called periods) and 18 vertical columns (called groups). (See appendix on page 180.) The properties of the elements in a particular row vary across it, thereby providing the concept of periodicity. Scientists refer to the elements contained in a particular vertical column as a group or family.

There are several versions of the periodic table used in modern science. The International Union of Pure and Applied Chemistry (IUPAC) recommends labeling the vertical columns from 1 to 18, starting with hydrogen (H) at the top of group 1 and ending with helium (He) at the top of group 18. The IUPAC further recommends labeling the periods (rows) from 1 to 7. Hydrogen (H) and helium (He) are the only two elements found in period (row) 1. Period 7 starts with francium (Fr) and includes the actinide series as well as the transactinides (very short-lived, human-made, superheavy elements).

The row (or period) in which an element appears in the periodic table tells scientists how many electron shells an atom of that particular element possesses. The column (or group) lets scientists known how many electrons to expect in an element's outermost electron shell. Scientists call an electron residing in an atom's outermost shell a *valence electron*. Chemists have learned that it is these valence electrons that determine the chemistry of a particular element. The periodic table is structured such that all the elements in the same column (group) have the same number of valence electrons. Consequently, the elements that appear in a particular column (group) display similar chemistry.

Chemists and physicists have found it useful to describe families of elements using such terms as *alkali metals, alkaline earth metals, transition metals, nonmetals,* and *noble gases*. A few of these groups (or families) of elements are summarized briefly here. For a more comprehensive discussion of the modern periodic table, see any contemporary college level chemistry textbook.

The *alkali metals* are the family of elements found in group 1 (column 1) of the periodic table. They include lithium (Li), sodium (Na), potassium (K), rubidium (Rb), cesium (Cs), and francium (Fr). These metals react vigorously with water to produce hydrogen gas. The *alkali earth metals* are the family of elements found in group 2 (column 2) of the periodic table. They include beryllium (Be), magnesium (Mg), calcium (Ca), strontium (Sr), barium (Ba), and radium (Ra). These elements are chemically reactive, but less so than the alkali metals.

The *noble gases* are the inert gaseous elements found in group 18 (column 18) of the periodic table. These elements include helium (He), neon (Ne), argon (Ar), krypton (Kr), xenon (Xe), and radon (Rn). The noble gases do not readily enter into chemical combination with other elements. Human-produced nuclear reactions result in the production of certain radioactive isotopes of krypton and xenon, while radon is a naturally occurring source of radioactivity.

The two long rows that appear below the periodic table are horizontal expansions corresponding to the positions of lanthanum (La) or actinium

MAKING UNUNTRIUM AND UNUNPENTIUM

In 2003, a team of American scientists from the Lawrence Livermore National Laboratory (LLNL), in collaboration with Russian scientists from the Joint Institute for Nuclear Research (JINR) in Dubna, Russia, discovered two new superheavy elements (SHEs). Scientists gave elements 113 and 115 the provisional names ununtrium (Uut) and ununpentium (Uup), respectively.

The successful experiments were conducted at the JINR U400 cyclotron with the Dubna gas-filled separator between July 14 and August 10, 2003. The

(continues)

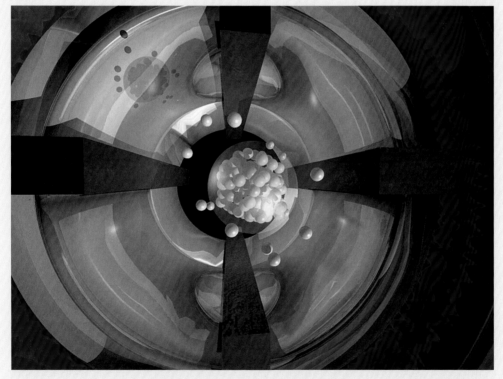

Computer-generated image depicting one of the numerous americium-243 target atoms with its nucleus of protons and neutrons surrounded by an electron cloud. In February 2004, scientists from Russia and the United States successfully bombarded americium-243 atoms with calcium-48 ions and ultimately produced several atoms of the short-lived, superheavy element ununtrium (Uut) with an atomic number of 113. *(DOE/LLNL)*

(continued)

scientific team observed nuclear decay chains that confirmed the existence of element 115 (ununpentium) and element 113 (ununtrium). Specifically, element 113 formed as a result of the alpha decay of element 115. The experiments produced four atoms each of elements 115 and 113 through the bombardment of an americium-243 target by high-energy calcium-48 nuclei.

Scientists suggest that ununtrium's most stable radioisotope is Uut-284, which has a half-life of about 0.48 second. Ununtrium then undergoes alpha decay into roentgenium-280. The substance is anticipated to be a solid at room temperature, but its density is currently unknown. Scientists classify ununtrium as a metal.

Ununpentium's most stable radioisotope is Uup-288, which has a half-life of about 87 milliseconds and then experiences alpha decay into Uut-284. Scientists classify ununpentium as a metal of currently unknown density. They also anticipate it would be a solid at room temperature.

(Ac). The lanthanoid (formerly lanthanide) series contains 15 elements that start with lanthanum (La, atomic number 57) and continue through lutetium (Lu, 71). All the elements in the lanthanoid series closely resemble the element lanthanum. Scientists often collectively refer to the 15 elements in the lanthanoid series along with the elements scandium (Sc) and yttrium (Y) as the rare earth elements or the rare earth metals. The *actinoid* (formerly actinide) *series* of heavy metallic elements starts with element 89 (actinium) and continues through element 103 (lawrencium). These elements are all radioactive and together occupy one position in the periodic table. The actinoid series includes uranium (U, 92) and all the human-made transuranium elements.

A transuranium (or transuranic) element is an element with an atomic number greater than that of uranium. These elements extend from neptunium (Np, atomic number 93) to lawrencium (Lr, 103). All transuranium elements are essentially human-made and radioactive, although scientists have detected extremely small trace amounts of natural neptunium and plutonium here on Earth. Some of the more significant transuranium elements are neptunium (93), plutonium (94), americium (95), curium (96), berkelium (97), and californium (98).

Scientists sometimes include in their discussion of transuranium elements the collection of human-made, superheavy elements (SHEs) that start with element number 104 (called rutherfordium [Rf]) and currently extend to element 118. These superheavy elements are also known as the *transactinide elements* because they appear in row (period) 7 of the periodic table after the element actinium (Ac). Fleeting amounts (one to several atoms) of all these elements have been created in complex laboratory experiments that involve the bombardment of special transuranium target materials with high-velocity nuclei.

CHEMICAL BONDS

Chemical energy is liberated or absorbed during a chemical reaction. In such a reaction, energy losses or gains usually involve only the outermost electrons of the atoms or ions of the system undergoing change; here, a chemical bond of some type is established or broken without disrupting the original atomic or ionic identities of the constituents. Scientists discovered that the atoms in many solids are arranged in a regular, repetitive fashion and that the atoms in this structured array are held together by interatomic forces called chemical bonds. Chemical bonds play a significant role in determining the properties of a substance. There are several different types of chemical bonds. These include *ionic bonds, covalent bonds, metallic bonds,* and *hydrogen bonds.*

Scientists define an ionic bond as one created by the electrostatic attraction between ions that have opposite electric charges. Sodium chloride (NaCl), better known as table salt, is a common example of ionic bonding. Sodium (Na) is a silvery metal that has one valence electron to lose, forming the cation (positive ion) Na^+. Chlorine (Cl) is a pale, yellow-green gas that has seven valence electrons and readily accepts another electron to form the anion (negative ion) Cl^-. The bond formed in creating a sodium chloride molecule is simply electrostatic attraction between sodium and chlorine. When a large number of NaCl molecules gather together, they form an ionic solid that has a regular crystalline structure. The ionic bond discussed here is the result of the sodium atom transferring a valence electron to the valence shell of a chlorine atom, forming the ions Na^+ and Cl^- in the process. Every Na^+ ion within the salt (NaCl) crystal is surrounded by six Cl^- ions, and every Cl^- ion by six Na^+ ions.

Scientists call the second type of chemical bond a covalent bond. In the covalent bond, two atoms share outer shell (valence) electrons. The molecular linkage takes place because the shared electrons are attracted to the positively charged nuclei (cores) of both atoms. One example is the hydrogen molecule (H_2), in which two hydrogen atoms are held tightly together. In 1916, the American chemist Gilbert Newton Lewis (1875–1946) proposed that the attractive force between two atoms in a molecule was the result of electron pair bonding—more commonly called covalent bonding. During the following decade, developments in quantum mechanics allowed other scientists to quantitatively explain covalent bonding. In most molecules, the atoms are linked by covalent bonds. Generally, in most molecules hydrogen forms one covalent bond; oxygen, two covalent bonds; nitrogen, three covalent bonds; and carbon, four covalent bonds. In the water molecule, the oxygen atom bonds with two hydrogen atoms. The H-O bonds form as the oxygen atom shares two pairs of electrons, while each hydrogen atom shares only one pair of electrons, which scientists refer to as *polar covalent bonds.*

In *metallic bonding,* electrons are freely distributed, so that many metal atoms share them. Immersed in this sea of negative electrons, the positive metal ions form a regular crystalline structure. Consider the metal sodium (Na) as an example. This metal is made up of individual sodium atoms, each of which has 11 electrons. In metallic bonding, every sodium atom releases one valence electron, which then moves throughout the metallic crystal, attracted to the positive Na^+ ions. It is this attraction that holds the metallic crystal together.

Finally, scientists describe *hydrogen bonding* as a weak to moderate attractive force due to polarization phenomena. This type of bonding occurs when a hydrogen atom that is covalently bonded in one molecule is at the same time attracted to a nonmetal atom in a neighboring atom. Hydrogen bonding typically involves small atoms characterized by high *electronegativity,* such as oxygen, nitrogen, and fluorine. Chemists define electronegativity as the attraction of an atom in a compound for a pair of shared electrons in a chemical bond. Under certain conditions during physical state transitions, hydrogen bonding occurs in the water molecule. Hydrogen bonds are also common in most biological substances.

CHAPTER 7

Discovering the Radioactive Nature of Matter

This chapter discusses how the discovery of radioactivity led to the modern understanding of matter at the atomic and subatomic levels. Wilhelm Conrad Roentgen started the ball rolling in 1895 with the discovery of X-rays. Understanding radioactivity made possible many other important but previously unimaginable developments. Breakthroughs included improved knowledge of the structure of matter through X-ray crystallography, new ways of exploring the cosmos through high-energy astrophysics, and radioactive tracers in medicine, industry, research, and industry.

X-RAYS LEAD THE WAY

The German experimental physicist Wilhelm Conrad Roentgen initiated both modern physics and the practice of radiology in 1895 with his world-changing discovery of X-rays. His breakthrough occurred at the University of Würzburg on the evening of November 8, 1895. Like many other late 19th-century physicists, Roentgen was investigating luminescence phenomena associated with cathode-ray tubes. His experimental device, also called a Crookes tube after its inventor, Sir William Crookes (1832–1919), consisted of an evacuated glass tube containing two electrodes, a cathode and an anode. Electrons emitted by the cathode often missed the anode and struck the glass wall of the tube, causing it to glow, or fluoresce.

On that particular evening, Roentgen decided to place a partially evacuated discharge tube inside a thick black cardboard carton. As he darkened the room and operated the light-shielded tube, he suddenly noticed that a paper plate covered on one side with barium platinocyanide began to fluoresce, even though it was located some 6.6 ft (2 m) from the discharge tube. He concluded that the phenomenon causing the sheet to glow was some new form of penetrating radiation that originated within the opaque paper-enshrouded discharge tube. He called the unknown radiation X-rays, because *x* is the traditional algebraic symbol for an unknown quantity.

During subsequent experiments, Roentgen discovered that objects of different thickness, when placed in the path of these mysterious X-rays, demonstrated variable transparency. He recorded the passage of X-rays through matter on photographic plates. As part of his investigations, Roentgen even held his wife's hand over a photographic plate and produced the first medical X-ray of the human body. When he developed this particular X-ray–exposed photographic plate, he observed an interior image of his wife's hand. It portrayed dark shadows cast by the bones within her hand and by the wedding ring she was wearing. These dark shadows were surrounded by a less darkened (penumbral) shadow corresponding to the fleshy portions of her hand. Roentgen formally announced this important finding on December 28, 1895. The discovery revolutionized the fields of physics, materials science, and medicine.

Although the precise physical nature of X-rays, as very short-wavelength, high-frequency photons of electromagnetic radiation, was not recognized until about 1912, the fields of physics and medicine immediately embraced Roentgen's discovery. Science historians regard this event as the beginning of modern physics. In 1896, the American inventor Thomas A. Edison (1847–1931) developed the first practical fluoroscope, a noninvasive device that used X-rays to allow a physician to observe how internal organs of the body function within a living patient. At the time, no one recognized the potential health hazards associated with exposure to excessive quantities of X-rays or other forms of ionizing radiation. Consequently, Roentgen, his assistant, and many early X-ray technicians eventually exhibited the symptoms of and suffered from acute radiation syndrome.

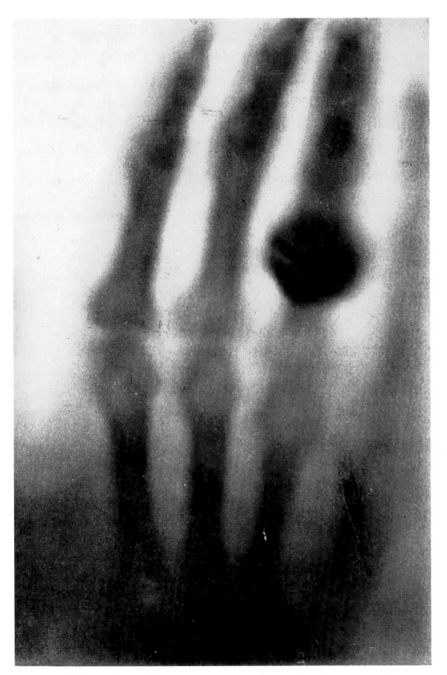

This is the first medical X-ray ever made. Wilhelm Conrad Roentgen made this image in late 1895, revealing the bones inside his wife's hand. The darker object is her wedding ring.   (U.S. National Library of Medicine/NIH)

Roentgen received numerous honors for the discovery of X-rays, including the first Nobel Prize in physics in 1901. Both a dedicated scientist and a humanitarian, Roentgen elected not to patent his discovery so the world could freely benefit from his work. He even donated his Nobel Prize money to the University of Würzburg. He died on February 10, 1923, in Munich, Germany.

X-RAYS AND MATERIALS SCIENCE

In 1912, the German physicist Max von Laue (1879–1960) postulated that the atoms in a crystalline solid might serve as a natural three-dimensional diffraction grating for X-rays. Pursuing this innovative idea, he developed an experiment in which X-rays were sent into a copper sulfate ($CuSO_4$) crystal. His technical assistants, Paul Knipping and Walter Friedrich, obtained a pattern of the diffracted X-rays on a photographic plate. Von Laue's scientific hunch was correct and led to the important field of X-ray crystallography. He received the 1914 Nobel Prize in physics.

Von Laue's experiment had two major impacts on physics and materials science. First, scientists could use X-rays of a known energy (and, therefore, a known wavelength) to investigate the atomic structure of crystals and other interesting materials, such as giant molecules. Second, scientists could use crystals of known physical structure to measure the wavelength (and, therefore, the energy) of incident X-rays.

Inspired by von Laue's discovery of X-ray diffraction, the British physicist Henry Gwyn Jeffreys Moseley (1887–1915) started to use the phenomenon in 1913 and observed a systematic relationship between the characteristic X-ray frequency (or wavelength) and the atomic number of the target material. Moseley produced characteristic X-rays for the approximately 40 different chemical elements he used as targets in electron bombardment experiments. He then plotted the frequencies of the characteristic X-ray spectra versus each element's atomic number. The linear pattern he observed led to the physical relationship that scientists call Moseley's law.

As a result of Moseley's brilliant work, scientists around the world began using a substance's characteristic X-ray spectrum as a measurable way of determining its atomic number. Moseley's law and his experimental approach showed scientists where elements were missing in the periodic table. In 1913, there were a number of empty places in the periodic table. One especially troublesome region of the periodic table involved the lanthanoid series.

At the time, the arrangement of the lanthanoid elements was somewhat imperfect because scientists were having a very difficult time separating and identifying them due to their very similar chemical properties. Once Moseley's work appeared, these gaps in the periodic table were quickly filled. In 1923, the Dutch physicist Dirk Coster (1889–1950) and the Hungarian radiochemist George de Hevesy (1885–1966) took advantage of hafnium's characteristic X-ray spectrum to identify that new element. Hafnium (Hf) is a rare earth element, that is chemically almost identical to zirconium (Zr). They named the new element after Niels Bohr's hometown of Copenhagen, Denmark (Latin name: Hafnia).

Most science historians believe that Moseley's substantial contributions to the knowledge of atomic structure would have qualified him for a Nobel Prize in physics. Unfortunately, the brilliant young physicist was killed by a sniper's bullet on August 15, 1915, while he was serving with the British army during the Battle of Gallipoli (Turkey) in World War I.

The phenomenon of X-ray diffraction gave rise to the science of X-ray crystallography. A father and son team of British scientists, Sir William Henry Bragg (1862–1942) and Sir William Lawrence Bragg (1890–1971), shared the 1915 Nobel Prize in physics for their pioneering efforts in the analysis of crystal structures by means of X-ray diffraction. The Braggs scattered X-rays of the proper wavelength off an atom's electron cloud. They then analyzed the resulting diffraction pattern to extract a great deal of information about the target material, including a three-dimensional portrait of the electron density within the crystal. Since many inorganic and organic substances form crystals, today's scientists employ computer-assisted X-ray crystallography to characterize the atomic structures of new materials.

Another British physicist, Charles Glover Barkla (1877–1944), devoted 40 years of his professional life to the detailed study of X-rays and how they interact with matter. He was awarded the 1917 Nobel Prize in physics for his discovery and detailed investigation of the characteristic X-ray radiation of the elements. Barkla recognized two fundamental types of characteristic X-rays, which he called K radiation (more penetrating) and L radiation (less penetrating). His work helped other scientists, such as the Swedish physicist Karl Manne Georg Siegbahn (1886–1978), expand

and refine the use of X-rays in classifying the structure of matter. Siegbahn was awarded the 1924 Nobel Prize in physics, which he received in 1925, for his contributions to X-ray spectroscopy.

HOW X-RAYS HELP UNLOCK MATTER'S SECRETS

When scientists bombard a solid target, such as molybdenum (Mo) or tungsten (W), with energetic electrons (typically in the kiloelectron volt [keV] range), X-rays are emitted from the material being bombarded. The accompanying figure shows the spectrum of X-rays produced when 35-keV electrons strike a molybdenum (Mo) target. This plot depicts relative X-ray intensity versus emitted X-ray wavelength (in nanometers [nm]). The two distinctive, high-intensity peaks (labeled K_α and K_β) represent the element's characteristic

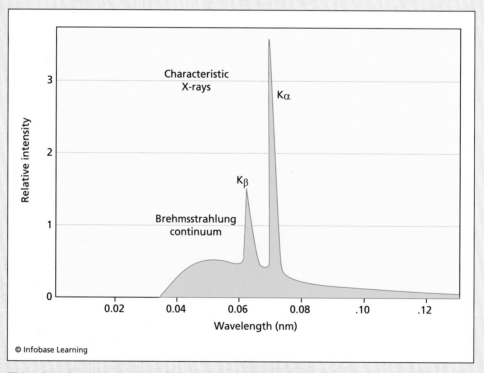

© Infobase Learning

This plot shows the spectrum of X-rays produced when 35-keV electrons strike a molybdenum (Mo) target. The two distinctive sharp peaks (called K_α and K_β) are the element's characteristic X-rays, while the smooth, lower-intensity continuous spectrum is due to brehmsstrahlung radiation. *(Author, based on DOE data)*

In the early 1930s, the British chemist Dorothy Mary Crowfoot Hodg-kin (1910–94) began doctoral work at Cambridge University. Her research efforts and subsequent professional activities led to the founding of pro-

X-rays, while the smooth, lower-intensity continuous spectrum is associated with brehmsstrahlung radiation. *Brehmsstrahlung* is the German word for braking.

When electrons that possess a minimum kinetic energy pass near an atomic nucleus in the target material, they decelerate rapidly (that is, they brake), and an X-ray photon is emitted as the incident electron loses energy. There is a prominent feature of this continuous spectrum, which scientists define as the *cutoff wavelength* (λ_{min}). Below the cutoff wavelength, the continuous (or brehmsstrahlung) spectrum does not exist. As early researchers discovered, the cutoff wavelength is totally dependent on the kinetic energy of the incident electron and completely independent of the target material. Should scientists decide to bombard a tungsten target with 35-keV electrons, all the features of tungsten's X-ray spectrum would be different—except for the cutoff wave-length (λ_{min}). In this example (involving 35-keV electrons), the cutoff wavelength is approximately 0.036 nanometer (nm).

The characteristic X-ray spectrum for the target material is of great use to scientists. The two distinctive sharp peaks (called K_α and K_β) arise as a result of a two-part physical process. When an incident energetic electron strikes one of the target atom's deep-lying electrons, that electron is kicked out of its shell (orbital position), and the incident electron is scattered. An electron in one of the higher energy shells of the target atom then jumps down to fill the deeper-lying vacancy (or hole) and emits its excess energy in the form of a character-istic X-ray. For historic reasons, scientists call the deep-lying electrons in the first ($n = 1$) shell K-shell electrons. If a higher energy L-shell ($n = 2$) electron fills the hole in the K shell, the transitioning electron emits an X-ray corresponding to the characteristic K_α line. If an M-shell ($n = 3$) electron jumps down to fill the K-shell vacancy, the transition electron emits an X-ray corresponding to the characteristic K_β line. An electron from another more energetic shell then fills the vacancy in the L or M shell.

Characteristic X-ray spectra are a bit more complicated than as discussed here, but the basic principle underlying the application of this important physi-cal phenomenon is elegantly simple. Scientists around the world routinely use characteristic X-ray spectra to identify substances.

tein crystallography. She successfully applied X-ray diffraction techniques to the detailed study of biological substances. Among her contributions to science were the molecular structures of cholesterol, penicillin, vitamin B-12, and insulin. She received the Nobel Prize in chemistry in 1964 for determining the structures of biochemical substances, including vitamin B-12, by X-ray techniques.

X-RAYS AND QUANTUM MECHANICS

The American physicist Arthur Holly Compton (1892–1962) performed the watershed X-ray scattering experiment that placed all of quantum mechanics on a sound experimental basis. Compton began his interest in X-rays as part of his doctoral research at Princeton University. In that effort, he studied the angular distribution of X-rays when deflected by crystals.

While working with Rutherford at the Cavendish Laboratory in England, Compton was able to verify the puzzling results obtained by other physicists—namely, that when scattered by matter, X-rays and gamma rays display an increase in wavelength as a function of scattering angle. Classical physics could not satisfactorily explain this phenomenon.

After a year of study in Great Britain, Compton returned to the United States and accepted a position as head of the department of physics at Washington University in Saint Louis, Missouri. There, using X-rays to bombard graphite (carbon), he resumed his investigation of the puzzling mystery of photon scattering and wavelength change. By 1922, Compton's experiments revealed that X-ray wavelength did increase with increasing scattering angle, a phenomenon scientists now call the *Compton effect*. He applied special relativity and quantum mechanics to explain the results and presented his famous quantum hypothesis in the 1923 paper "A Quantum Theory of the Scattering of X-rays by Light Elements."

In 1927, Compton shared the Nobel Prize in physics with the Scottish scientist Charles Wilson (1869–1959). Compton caused a revolution in physics with his seemingly simple assumption that X-ray photons behaved like particles and interacted one-on-one as they scattered with free electrons in the target material. It was Wilson's Nobel Prize–winning cloud chamber that helped verify the presence of the recoil electrons predicted by Compton's quantum scattering hypothesis. Telltale cloud tracks of recoiling electrons provided indisputable evidence of the particlelike behavior of electromagnetic radiation. Compton's pioneering research implied that the scattered X-ray photons had less energy than the original X-ray photons. Compton was the first scientist to experimentally demonstrate the quantum (or particlelike) nature of electromagnetic waves. The discovery of the Compton effect served as the technical catalyst for the acceptance and further development of quantum mechanics in the 1920s and 1930s.

THE DISCOVERY OF RADIOACTIVITY

While exploring the phosphorescence of uranium salts in early 1896, the French physicist Antoine-Henri Becquerel (1852–1908) accidentally stumbled upon the phenomenon of radioactivity.

Roentgen's discovery of X-rays in late 1895 encouraged Becquerel to investigate whether there was any connection between X-rays and naturally occurring phosphorescence in uranium salts. Becquerel began to experiment by exposing various crystals to sunlight and then placing each crystal on an unexposed photographic plate that was wrapped in black paper. As a trained scientist, he reasoned that if any X-rays were so produced, they would penetrate the wrapping and create a characteristic spot that would be developed on the photographic plate. Following this line of experimentation with various uranium salts, it was quite by accident that he stumbled upon the actual phenomenon of radioactivity.

For several days it was cloudy in Paris, so he could not expose the uranium salts to direct sunlight. Inadvertently, he placed a piece of uranium salt (one that had not been exposed to sunlight) on top of a dark paper–wrapped photographic plate. Because of the inclement weather, he kept the combination in a drawer. Then, for no particular reason on March 1, Becquerel decided to develop this particular black paper–wrapped photographic plate. To his surprise, even though the uranium salt was not fluorescing due to exposure to the ultraviolet radiation found in sunlight, the crystalline material still produced an intense silhouette of itself on the photographic plate. As a trained scientist, Becquerel properly concluded that some new type of invisible rays, perhaps similar to Roentgen's X-rays, were emanating from the uranium compound. He also recognized that sunlight had nothing to do with this penetrating new phenomenon.

Becquerel announced the finding in a short paper that he read to the French Academy of Science on March 2, 1896. This new phenomenon remained known as Becquerel rays until another nuclear science pioneer, Marie Curie (1867–1934), called the phenomenon *radioactivity* in 1898.

Becquerel shared the 1903 Nobel Prize in physics with Pierre (1859–1906) and Marie Curie for his role in the discovery of radioactivity. In his honor, scientists named the SI unit of radioactivity the *becquerel* (Bq). One becquerel represents one disintegration (decay) per second.

MARIE CURIE

Collaborating with her professor husband (Pierre), the Polish-French radiochemist Marie Curie isolated and identified the radioactive elements radium (Ra) and polonium (Po) in 1898. Their achievements represent one of the great milestones in materials science.

Marie Curie was born Maria Sklodowska on November 7, 1867, in Warsaw, Poland. In 1891, she joined her older sister, Bronya, who was living in Paris. There, three years later (in 1894), Maria Sklodowska met Pierre Curie, an accomplished physicist who was the head of a laboratory at the School of Industrial Physics and Chemistry. They married in July 1895. At the wedding, the gifted Polish graduate student adopted France as her new home and became known as Marie Curie.

While pregnant with her first child in 1897, Marie Curie began searching for an interesting doctoral research topic. Her husband suggested that she consider investigating the mysterious phenomenon recently reported by Becquerel. Intrigued with the prospect, Marie started one of the most important research legacies in the history of science. The effort was slightly delayed due to the birth of the couple's first daughter, Irène, in September 1897.

Becquerel's announcement to the French Academy in March 1896 had not caused much of a scientific stir, but Marie Curie began to vigorously investigate the new phenomenon in early 1898 and soon coined the word *radioactivity* to describe it. She quickly discovered that, much like uranium, thorium also emitted so-called Becquerel rays.

First, she confirmed Becquerel's original findings, then she found that two uranium minerals, pitchblende and chalcolite, were actually more active than uranium itself. In early 1898, she also came to the very important conclusion that these uranium ores must actually contain other more intensely radioactive elements. Assisted by Pierre, she began the arduous search for these suspected new radioactive substances. Marie methodically ground up and chemically processed tons of pitchblende in small 44-lbm (20-kg) batches. This tedious extraction effort slowly began to yield very minute but chemically identifiable quantities of the elements polonium and radium. The Curies discovered polonium in July 1898 and named the naturally radioactive element after Marie's native land. They announced the discovery of radium in September 1898. The French chemist Eugène Demarçay (1852–1904) used spectroscopy to help confirm the presence of radium, a naturally occurring radioactive element that is chemically similar to barium. By early 1902, the Curies had obtained about one-tenth of a gram of radium chloride. Their primitive processing laboratory was set up in a vacant shed, and its contents soon began to exhibit the glow of a faint blue light, the sign of extensive radioactive contamination due to their monumental extraction efforts.

This 1967 French stamp honors the birth centenary of Marie Curie. Included in the stamp's design is a glowing bowl of radium—the naturally occurring radioactive element that she codiscovered with her husband (Pierre) in September 1898. *(Author)*

Marie Curie successfully presented her doctoral dissertation in 1903 and became the first woman to earn an advanced scientific degree in France. That same year, she also became the first woman to win the Nobel Prize when she shared the physics award with her husband, Pierre, and Becquerel for their pioneering research efforts involving radioactivity.

Tragedy struck on April 19, 1906, when Pierre Curie was killed in a Parisian street accident. The accident left Marie Curie a widow with two young daughters, Irène, aged 9, and Ève, aged 2. She rejected the offer of a modest pension and decided to support her family by filling her husband's teaching position as a professor of physics at the Sorbonne. By 1908, she finally overcame intense gender bias and began serving the Sorbonne as its first female professor.

She received the 1911 Nobel Prize in chemistry in recognition of her work involving the discovery and investigation of radium and polonium. That award made her the first scientist to win a second Nobel Prize. Her research demonstrated that one element could transmute into another through the process of radioactivity. The concept revolutionized chemistry and materials science.

Marie Curie founded the Radium Institute of the University of Paris in 1914 and became its first director. Throughout her life, she actively promoted the medical applications of radium. During World War I, she and her daughter Irène, a future Nobel laureate, contributed to the French war effort by training young women in X-ray technology. They also assisted physicians by operating radiology equipment under battlefield conditions. These patriotic efforts exposed them both to large doses of ionizing radiation.

As a lasting international tribute to Pierre and Marie Curie, fellow scientists named the traditional unit of radioactivity the *curie*. The transuranium element curium (Cm) also honors their contributions to science. Madam Curie died of leukemia on July 4, 1934, in Savoy, France.

MEASURING THE RADIOACTIVITY OF MATTER

For many atoms, the protons and neutrons arrange themselves in such a way in the nucleus that they become unstable and spontaneously disintegrate at different but statistically predictable rates. As an unstable nucleus attempts to reach stability, it emits various types of nuclear radiation. Marie Curie and Ernest Rutherford called the first three forms of nuclear radiation alpha (α) rays, beta (β) rays, and gamma (γ) rays. They assigned the letters of the Greek alphabet to these phenomena in the order of their discovery. Marie Curie prepared a diagram in her doctoral dissertation that described how alpha rays bent one way in a magnetic field, beta rays bent the opposite way in the same magnetic field, and gamma rays appeared totally unaffected by the presence of a magnetic field. At the time, the precise nature of the interesting rays that emanated from naturally radioactive substances was not completely understood.

Scientists use the general term *radiation* to describe the propagation of waves and particles through the vacuum of space. The term includes both electromagnetic radiation (EMR) and nuclear particle radiation. Electromagnetic radiation has a broad continuous spectrum that includes radio waves, microwaves, infrared radiation, visible radiation (light), ultraviolet radiation, X-rays, and gamma rays. (See chapter 5.) Photons of EMR travel at the speed of light. With the shortest wavelength (λ) and highest frequency (ν), the gamma ray photon is the most energetic. By way of comparison, radio wave photons have energies between 10^{-10} and 10^{-3} electron volt (eV), visible light photons between 1.5 and 3.0 eV, and gamma ray photons between approximately 1 and 10 million electron volts (MeV).

One of the defining characteristics of radiation is energy. For ionizing radiation, a common unit is the electron volt (eV), the kinetic energy that a free electron acquires when it accelerates across an electric potential difference of one volt. The passage of ionizing radiation through matter causes the ejection of outer (bound) electrons from their parent atoms. An ejected (or free) electron speeds off with its negative charge and leaves behind the parent atom with a positive electric charge. Scientists call the two charged entities an *ion pair*. On average, it takes about 25 eV to produce an ion pair in water. By rapidly creating a large number of ion pairs in a small volume of matter, ionizing radiation can damage or harm many material substances, including living tissue.

The term *nuclear radiation* refers to the particles and electromagnetic radiations emitted from atomic nuclei as a result of nuclear reaction processes, such as radioactive decay, fission, fusion, and matter-antimatter annihilation. The most commonly encountered forms of nuclear radiation are alpha particles, beta particles, gamma rays, and neutrons, but nuclear radiation also appears in other forms, such as energetic protons from accelerator-induced reactions. When discussing the biological effects of nuclear radiation, scientists often include X-rays in the list of ionizing radiations. Although energetic photons, X-rays are not (as commonly misunderstood) a form of nuclear radiation, because they do not originate from processes within the atomic nucleus.

Radioactivity is the spontaneous decay or disintegration of an unstable nucleus, and the emission of nuclear radiation usually accompanies the decay process. Scientists measure how radioactive a substance is with either of two basic units of radioactivity: the curie (Ci) and the becquerel (Bq). The curie is the traditional unit used by scientists to describe the intensity of radioactivity in a sample of material. One curie (Ci) of radioactivity is equal to the disintegration or transformation of 37 billion (37 × 10^9) nuclei per second. This unit honors Marie and Pierre Curie, who discovered radium in 1898. The curie corresponds to the approximate radioactivity level of one gram of pure radium—a convenient, naturally occurring radioactivity standard that arose in the early days of nuclear science. The becquerel is the SI unit of radioactivity. One becquerel (Bq) corresponds to the disintegration (or spontaneous nuclear transformation) of one atom per second.

When the nuclei of some atoms experience radioactive decay, a single nuclear particle or gamma ray appears, but when other radioactive nuclides decay, a nuclear particle and one or more gamma rays may appear simultaneously. Scientists recognize that the curie or the becquerel describe radioactivity in terms of the number atomic nuclei that transform or disintegrate per unit of time and not necessarily the corresponding number of nuclear particles or photons emitted per unit of time. To obtain the latter, they usually need a little more information about the particular radioactive decay process.

Each time an atom of the radioisotope cobalt-60 decays, for example, its nucleus emits an energetic beta particle (with approximately 0.31 MeV energy) along with two gamma ray photons—one at an energy of 1.17 MeV and the other at an energy of 1.33 MeV. During this decay process, the radioactive cobalt-60 atom transforms into the stable nickel-60 atom.

Scientists cannot determine the exact time that a particular nucleus within a collection of the same type of radioactive atoms will decay. As first quantified by Ernest Rutherford and Frederick Soddy (1877–1965) in the early 1900s at McGill University in Canada, the average behavior of the nuclei in a very large sample of a particular radioisotope can be predicted quite accurately using statistical methods. Their findings became known as the law of radioactive decay. The concept of half-life ($T_{1/2}$) is a very important part of this physical law. The radiological half-life is the period of time required for half of the atoms of a particular radioactive isotope to disintegrate or decay into another nuclear form. The half-life is a characteristic property of each type of radioactive nuclide. Experimentally measured half-lives vary from millionths of a second to billions of years.

A decay chain is the series of nuclear transitions or disintegrations that certain families of radioisotopes undergo before reaching a stable end nuclide. Following the tradition introduced by early nuclear scientists, scientists call the lead radioisotope in a decay chain the *parent nuclide* and subsequent radioactive isotopes in the same chain the *daughter nuclides,* granddaughter nuclides, great-granddaughters, and so forth until they reach the last stable isotope for that chain. As shown in the accompanying figure, the decay chain for some naturally occurring radioisotopes (such as radium-226) are temporally dynamic and chemically complex. Radium-226 is actually part of a much larger natural decay chain called the uranium-238 decay series. That decay series starts with the element uranium's most common isotope, uranium-238, as the parent radioisotope (half-life 4.5×10^9 years) and ends with the stable isotope of the element lead (Pb-206).

The decay constant (λ) is related to the half-life by the mathematical expression $\lambda = (\ln 2)/T_{1/2} \approx 0.693/T_{1/2}$. When scientists say that the important medical and industrial radioisotope cobalt-60 ($^{60}_{27}$Co) has a half-life of

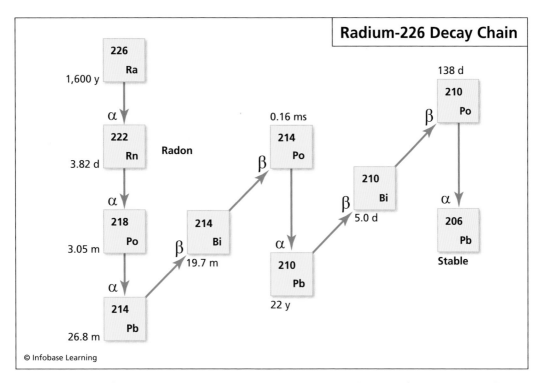

Radium-226 Decay Chain

The decay chain for the naturally occurring radioisotope radium-226 (1,600 years half-life). The half-lives for each of the members of this decay chain are provided as well as the specific radioactive decay modes, namely alpha (α) decay and beta (β) decay. Note how radioactive decay transforms the unstable parent element into a variety of unstable intermediate elements (including an unstable isotope of lead [Pb-214]) before ending up as a stable form of lead (the isotope Pb-206). *(NIST)*

5.27 years, they also imply that the decay constant for this radioisotope is $\lambda_{cobalt-60} = 0.1315$ years^{-1}. The decay constant is a physical property of matter and represents the fractional change in radioactive nuclide concentration per unit of time. Stable isotopes have a decay constant of zero. The decay constant is very useful in understanding the intensity of radioactivity and the dynamic behavior of radioisotope populations. At a particular moment in time, scientists can express the activity (or radioactivity) of a given quantity of radioactive substance (N) as A (activity) = N λ. Scientists note that the activity (A) has the units of decays or disintegrations per unit time. As previously mentioned, the two basic units of activity are the curie and the becquerel.

MATTER CONSISTS OF ISOTOPES

At the start of the 20th century, the British chemist Frederick Soddy collaborated with Ernest Rutherford in developing the law of radioactive decay, a scientific principle of great importance. They also performed an early investigation of the gaseous emanation of radium. About a decade later, Soddy independently developed the concept of the isotope.

By 1903, Soddy had departed from Canada, returned to Great Britain, and accepted a position working with the Scottish chemist Sir William Ramsay (1852–1916) at University College in London. They used spectroscopic techniques to demonstrate that the element helium is released during the radioactive decay of radium. Their research made an important connection—namely, that the alpha particle emitted during the radioactive decay of radium (and other heavy nuclei) is really the nucleus of a helium atom.

From 1904 to 1914, Soddy was a lecturer in physical chemistry and radioactivity at the University of Glasgow. He evolved a so-called displacement law by suggesting that the emission of an alpha particle from an element during radioactive decay causes that element to move back

TRACER PRINCIPLE

Nuclear medicine employs very small amounts of radioactive materials to diagnose and treat diseases. One of the fundamental tools of nuclear medicine is the tracer principle, discovered in 1913 by the Hungarian radiochemist George de Hevesy. While working in Ernest Rutherford's laboratory in Manchester, England, de Hevesy performed an unsuccessful experiment that nevertheless led to the very important conclusion that it is impossible to isolate (by chemical means) a radioisotope from the element of which it is a part. Since radioactive atoms remain the faithful companions of the nonradioactive atoms of the same element, he suggested that a radioactive isotope's characteristic nuclear radiation could serve as a special marker. This radioactive marker, or tracer, will follow its companion nonradioactive atoms through a living system as the isotopes of the particular element experience various biophysical and biochemical processes. Because radiation detectors can observe the presence of minuscule quantities of ionizing radiation, even a very tiny amount of radioactive material mixed in with nonradioactive atoms of the same element is sufficient as a tracer.

(or displace itself) two places in the periodic table. Pursuing this line of thinking further, he proposed in 1913 the concept of the *isotope,* a word derived from ancient Greek and meaning "same place." In a brilliant integration of available experimental data on radioactivity, he postulated that certain elements exist in two or more forms that have different atomic weights but nearly indistinguishable chemical characteristics.

Soddy's work allowed scientists to unravel the bewildering variety of new radioactive substances that had been discovered during the previous decade. The isotope concept made things fit together more logically on the periodic table of chemical elements. Soddy's work also prepared the scientific community for Rutherford's identification of the proton in 1919 and his subsequent speculation about the possible existence of a neutron. Soddy's isotope hypothesis became completely appreciated only after the British physicist James Chadwick (1891–1974) discovered the neutron in 1932.

Soddy received the 1921 Nobel Prize in chemistry for his research activities involving the origin and nature of isotopes. Although he had made some excellent contributions to chemistry and materials science

De Hevesy received the 1943 Nobel Prize in chemistry for developing this valuable concept.

Starting in the 1930s, the creation of artificial radioactive isotopes greatly expanded the application of de Hevesy's tracer principle in medicine, agriculture, science, and industry. Three particular radioisotope tracers, carbon-14, tritium (hydrogen-3), and phosphorous-32, played a major role in the establishment of modern biochemistry after World War II. These tracers are still important in modern biochemical research.

Similarly, the tracers carbon-11, iodine-123, fluorine-18, and technetium-99m are key tools for in vivo biochemistry research and nuclear medicine applications. In vivo procedures take place when health care workers give trace amounts of radiopharmaceuticals (medical radioisotopes) directly to a patient. A radiopharmaceutical is the basic radioactively tagged compound (tracer) used to produce a nuclear medicine image. By comparison, an in vitro procedure takes place outside a patient's body in a test tube. For example, radioimmunoassay (RIA) is a special type of in vitro procedure that combines the use of radioactive chemicals and antibodies to measure the levels of hormones, drugs, and vitamins in a patient's blood.

between 1900 and 1914, once he left the University of Glasgow, he basically abandoned any further technical contributions to these fields.

ASTON'S MASS SPECTROMETER

Isotopes are atoms of the same chemical element but with different numbers of neutrons in their nuclei. An isotope is specified by its relative atomic mass (A) and a symbol denoting the chemical element, such as "uranium-235" or "$^{235}_{92}U$" to represent the uranium atom that has 92 protons and 143 neutrons in its nucleus. Isotopes can be either stable or unstable (radioactive). Soddy based the original isotope hypothesis on his work with radioactive elements. The invention of the mass spectrometer by another British scientist, Francis W. Aston (1877–1945), expanded this important concept to include a large number of nonradioactive isotopes.

Aston was born on September 1, 1877, near Birmingham, Great Britain. In 1910, he joined the Cavendish Laboratory at Cambridge University and began working as an assistant to the Nobel laureate Sir Joseph John (J. J.) Thomson. At the time, Thomson was investigating methods of deflecting positively charged particles with magnetic fields and using electric fields (maintained at different voltages) to separate ions by their charges and masses. Aston improved the equipment being used by Thomson. Using neon (Ne) gas, Aston then started to investigate the fundamental principles underlying the operation of a mass spectrometer.

World War I interrupted Aston's initial efforts, but in 1919 he returned to Cambridge and constructed his first mass spectrometer (originally called a mass spectrograph). Over the next two years, he improved the design of the mass spectrometer and was eventually able to identify 212 naturally occurring, nonradioactive isotopes. His work produced a revolution in the understanding of matter and earned him the 1922 Nobel Prize in chemistry. Scientists around the world quickly recognized that the isotope concept was valid and that the elements generally consisted of several nonradioactive isotopes as well as possibly one or more radioactive isotopes.

Eventually scientists would learn, for example, that the element carbon (C) has a number of interesting and important isotopes. Carbon-12 and carbon-13 are stable isotopes. Carbon-11 is radioactive and decays into boron-11 by positron emission, while carbon-14 is radioactive and decays into nitrogen-14 by beta emission. A positron is a positively charged electron, while a beta particle is an ordinary negatively charged electron that emerges out of the nucleus during radioactive decay.

The mass spectrometer is an essential instrument in many fields of science. It can identify the relative abundances of isotopes and measure the masses of atoms and molecules. The basic mass spectrometer operates under the following general principles. The sample to be analyzed is vaporized, ionized, and injected into an evacuated chamber. An electric field accelerates the positively charged ions down the chamber, and a uniform magnetic field then separates the accelerated ions according to their mass-to-charge ratios. An appropriate detector then reports the fractional abundances of each of the ions present.

Aston began his investigations of nonradioactive isotopes using the element neon (Ne). As the ionized neon gas moved through his early spectrometers, the gas separated into three distinct fragments, corresponding to the nonradioactive isotopes neon-20 (0.9051 fractional abundance), neon-21 (0.0027 fractional abundance), and neon-22 (0.0922 fractional abundance).

Researchers can increase or decrease the speed of an ion by changing the voltage of the mass spectrometer's electric field. They can also alter the direction of an ion by changing the strength of the instrument's magnetic field. The mass-to-charge ratio of an ion determines the magnitude of its deflection as it travels through a particular mass spectrometer. According to the laws of physics, a lower-mass ion will experience more of a deflection than a higher-mass ion. Modern mass spectrometers are often designed to operate in either the positive ion mode or the negative ion mode. Scientists routinely use mass spectrometry to assess the isotopic composition of elements within a given substance.

CHAPTER 8

Exploring the Atomic Nucleus

This chapter shows how scientists explored the atomic nucleus and tapped the incredible amounts of energy hidden within. Scientific discoveries about the nuclear atom and companion technical developments forever changed the trajectory of human civilization by creating the nuclear age.

ERNEST RUTHERFORD—FATHER OF THE ATOMIC NUCLEUS

The New Zealander-British physicist Ernest Rutherford boldly introduced his concept of the nuclear atom in 1911. By assuming that each atom consisted of a small positively charged central region that was surrounded at great distance by orbiting electrons, Rutherford simultaneously transformed all previous scientific understanding of matter and established the world-changing discipline of nuclear physics.

Rutherford was born on August 30, 1871, near Nelson, New Zealand, and earned a scholarship in 1889 to Canterbury College at the University of New Zealand (Wellington). He graduated from that institution in 1893 with a double master of arts degree in mathematics and physical science. The following year, he received a bachelor of science degree.

In 1894, Rutherford traveled to Trinity College, Cambridge (England). He became a research student at the Cavendish Laboratory working under

Sir Joseph John (J. J.) Thomson. A skilled experimental physicist, Rutherford collaborated with Thomson on studies involving the ionization of gases exposed to X-rays. In 1898, he reported the existence of alpha rays and beta rays in uranium radiation. Rutherford created the basic language scientists still use to describe radioactivity, emanations from the atomic nucleus, and various constituents of the nucleus, such as the proton and the neutron. His discovery of positively charged alpha rays and negatively charged beta rays marked the beginning of his many important contributions to nuclear physics.

Rutherford went to Montreal, Canada, in 1898 to accept the chair of physics at McGill University. There, he collaborated with Frederick Soddy and produced many papers, including the one that presented the disintegration theory of radioactive decay. Rutherford proved to be an inspiring leader and steered many future Nobel laureates toward their great achievements in nuclear physics or chemistry. The German radiochemist Otto Hahn (1879–1968) worked under Rutherford at McGill from 1905 to 1906. Hahn discovered a number of new radioactive substances and was able to establish their positions in the series of radioactive transformations associated with uranium- and thorium-bearing minerals.

The complex radioactive decay phenomena exhibited by minerals containing uranium or thorium offered Rutherford a great challenge and a great research opportunity. In 1905, he published *Radio-Activity,* a book that influenced nuclear research for many years. Rutherford received the 1908 Nobel Prize in chemistry for his "investigations into the disintegrations of the elements and the chemistry of radioactive substances."

Rutherford returned to Great Britain in 1907 and became professor of physics at the University of Manchester. Starting in about 1909, two of his students began a series of alpha particle scattering experiments in Rutherford's Manchester laboratory. The insightful interpretation of data from the experiments allowed Rutherford to completely transform existing knowledge of the atom. Rutherford suggested that the British-New Zealander scientist Ernest Marsden (1888–1970) and the German physicist Hans Geiger (1882–1945) perform an alpha particle scattering experiment. Responding to their mentor, they bombarded a thin gold foil with alpha particles and diligently recorded the results. They were amazed when about one in 8,000 alpha particles bounced back in their direction. This simply should not occur under Thomson's plum pudding model of the atom. Thomson's atomic model assumed the atom was a uniform mixture of matter embedded with positive and negative charges much like raisins in a plum pudding. Rutherford quickly grasped the great significance of the experimental results. He later remarked that this amazing experiment was "As if you fired a 15-inch (38-cm) naval shell at a piece of tissue paper and the shell came right back and hit you."

Rutherford used the results of the gold foil alpha scattering experiment to postulate the existence of the atomic nucleus. In 1911, he introduced what physicists term the *Rutherford nuclear atom*. In this model, Rutherford proposed that the atom consisted of a small, positively charged central region (named the *nucleus*) surrounded at a great distance by electrons traveling in circular orbits, much like the planets in the solar system orbit the Sun. Rutherford's nuclear atom completely revised the scientific understanding of matter.

Originally conceived using classical physics, the Rutherford atom was actually unstable. Under Maxwell's classical electromagnetic theory, electrons traveling in circular orbits should emit electromagnetic radiation, thereby losing energy and eventually tumbling inward toward the nucleus. However, the brilliant Danish physicist Niels Bohr (1885–1962) arrived at Rutherford's laboratory in 1912 and eventually resolved this thorny theoretical dilemma. Bohr refined Rutherford's model of the nuclear atom by using Max Planck's quantum theory to assign specific nonradiating orbits to the electrons. (See next section.) The *Bohr atomic model* worked, and the Danish physicist received the 1922 Nobel Prize in physics for the concept. Nevertheless, it was Rutherford's original insight that completely transformed the scientific understanding of the atom and earned him the title "father of the atomic nucleus."

Following World War I, Rutherford returned to Manchester and made another major contribution to nuclear physics. By bombarding nitrogen with alpha particles, he achieved the first nuclear transmutation reaction. Rutherford's energetic alpha particles smashed into the target nitrogen nuclei, causing a nuclear reaction that produced oxygen and emitted a positively charged particle. Rutherford identified the emitted particle as the nucleus of the hydrogen atom and named it the *proton*. He speculated further that the nucleus might even contain a companion neutral particle, which he termed the *neutron*. One of Rutherford's research assistants, Sir James Chadwick, took up the quest for this elusive neutral nucleon and eventually discovered it in 1932.

In 1919, Rutherford became director of the Cavendish Laboratory at Cambridge University. He transformed the famous laboratory into a world center for nuclear research and mentored other future Nobel laureates, including Sir John Douglas Cockcroft (1897–1967), Ernest Walton (1903–95), and Chadwick. He died on October 19, 1937, in Cambridge. As a lasting tribute to the person who did so much to establish the field of nuclear physics, Rutherford's ashes were placed in the nave of Westminster Abbey, near the resting places of two other great scientists, Sir Isaac Newton and Lord Kelvin. In his honor, scientists named element 104 rutherfordium (Rf).

THE BOHR MODEL OF THE ATOM

Rutherford's nuclear atom model, while revolutionary, also posed a major scientific dilemma. Under classical electromagnetic theory, a charge (such as an orbiting electron) moving in a circular (or curved) orbit should lose energy. Rutherford's atomic nucleus model did not suggest any mechanism that would prevent an orbiting electron from losing its energy and falling into the positively charged nucleus under the influence of Coulomb attraction. (Under the laws of electrostatics, like electric charges repel and unlike charges attract each other.) Stable, electrically neutral atoms clearly existed in nature, so scientists began to question whether classical electromagnetic theory was wrong or whether Rutherford's nuclear atom model was incorrect. It took another revolution in thinking about the structure of the atom to resolve this problem.

In 1913, Niels Bohr combined Max Planck's quantum theory with Rutherford's nuclear atom model and produced the Bohr model of the hydrogen atom. Bohr's stroke of genius assumed that in the hydrogen atom, the electron could occupy only certain discrete energy levels, or orbits, as permitted by quantum mechanics. Scientists quickly embraced Bohr's new atomic model because it provided a plausible quantum mechanical explanation for the puzzling line spectra of atomic hydrogen. Science historians regard Bohr's work in atomic structure as the beginning of modern quantum mechanics.

Other talented physicists refined Bohr's model and produced acceptable quantum mechanical models for atoms more complicated than hydrogen. Famous contributors to the maturation of quantum mechanics included the German theoretical physicist Werner Heisenberg (1901–76) (the uncertainty principle) and the Austrian-Swiss theoretical physicist Wolfgang Pauli (1900–58) (the exclusion principle). As a result of these efforts, scientists began describing the ground state electronic configurations of the elements in the periodic table in terms of configurations (or

This 1963 Danish stamp honors Niels Bohr and his quantum theory of atomic structure. *(Author)*

shells) of electrons in certain allowed quantum mechanical energy states. The configuration of the outermost electrons of an atom determines its chemical properties.

A great deal of elegant atomic physics, both theoretical and experimental, allowed scientists to describe the cluster of electrons surrounding the nucleis of atoms beyond hydrogen in terms of a set of quantum numbers. The set of quantum numbers consisted of the principal quantum number *(n)*, the orbital quantum number *(l)*, the magnetic quantum number (m_l), and the spin quantum number (m_s). Although detailed discussion of this interesting work is beyond the scope of this chapter, a brief summary will illustrate some of the marvelous intellectual achievements that occurred in the two decades following the introduction of the Bohr atom.

The principal quantum number *(n)* is the quantum mechanical property upon which the energy of an electron in an atom mainly depends. Atomic scientists state that all orbiting electrons with the same principal quantum number *(n)* are in the same shell. For example, two electrons with $n = 1$ are in a single shell (traditionally called the K shell). Similarly, electrons with $n = 2$ are in another shell (referred to as the L shell). As the atomic number (Z) of a chemical increases, so do other electron shells, leading to much more complicated atomic structures. Scientists use the following sequence of electron shells: K $(n = 1)$, L $(n = 2)$, M $(n = 3)$, N $(n = 4)$, O $(n = 5)$, P $(n = 6)$, and Q $(n = 7)$.

The angular momentum quantum number *(l)* differentiates between electron orbital shapes. This quantum mechanical property can assume integer values from 0 to $n - 1$. Scientists find it convenient to treat orbiting electrons with the same value of *n* but different *l* as being in the same subshell. The K shell (for which $n = 1$) consists of a single subshell, while the L shell (corresponding to $n = 2$) has two subshells. For historic reasons, scientists refer to these subshells by lowercase letters rather than using the values of the appropriate orbital quantum number *(l)*. The $l = 0$ subshell is called the *s* subshell, while the $l = 1$ subshell is called the *p* subshell, and so forth. This notational approach creates the traditional subshell designation sequence *s* $(l = 0)$, *p* $(l = 1)$, *d* $(l = 2)$, *f* $(l = 3)$, *g* $(l = 4)$, *h* $(l = 5)$, and so on.

The magnetic quantum number (m_l) distinguishes between electrons that have the same energy and orbital shape (that is, the same *n* and *l* quantum numbers) but possess a different orientation in space. Finally, the spin quantum number (m_s) describes the two possible orientations an electron's spin axis can assume. To envision this quantum property, imagine a spinning electron acts like a very tiny bar magnet, with a north and south magnetic pole.

In the era of modern physics, scientists realized that every atom can possess an incredibly large number of electron positions. In quantum mechanics, the *ground state* means an atom with all its electrons in their lowest energy states. The *excited state* refers to an atom in which some or all its electrons reside in higher (than normal) energy states.

Chemists and physicists relate the chemical properties of an atom primarily with the electron configuration of the ground state. For the neutral carbon atom (which has six electrons), the ground state electron configuration is conveniently described by the following quantum mechanical shorthand: $1s^2 2s^2 2p^2$. Similarly, the sulfur atom (S) (which has 16 electrons) is described in the ground state by the following electronic configuration: $1s^2 2s^2 2p^6 3s^2 3p^4$.

CHADWICK DISCOVERS THE NEUTRON

By discovering the neutron in 1932, Sir James Chadwick enabled the development of modern nuclear technology. The neutron is the magic bullet that provides access to the energy within the atomic nucleus and supports the transmutation of chemical elements.

Chadwick was born on October 20, 1891, in Cheshire, England. Following graduation from college, he had the opportunity to explore various problems in radioactivity and the emerging field of nuclear science. From 1912 to 1913, Chadwick worked under Rutherford and received a master of science degree in physics. Later in 1913, he accepted a scholarship to work under Hans Geiger in Berlin. Unfortunately, shortly after Chadwick arrived in Germany, World War I broke out, and he was interned as an enemy alien for the next four years.

Chadwick returned to Great Britain following World War I and rejoined Rutherford, who by that time had become director of the Cavendish Laboratory at Cambridge University. At the Cavendish Laboratory, Chadwick collaborated with Rutherford in accomplishing the transmutation of light elements and in investigating the properties and structure of atomic nuclei. Tracking down and identifying the elusive neutron remained one of his major research objectives. From 1923 to 1935, Chadwick served as the assistant director of research for the Cavendish Laboratory. He assumed more and more responsibility for the operation of this laboratory as Rutherford advanced in age.

Chadwick's quest for the neutron eventually relied upon earlier experiments done in Germany and France, the results of which were mistakenly interpreted by each research team. In 1930, the German physicist Walter Bothe (1891–1957) bombarded beryllium (Be) with alpha particles and observed that the ensuing nuclear reaction gave off a strange, penetrating radiation that was electrically neutral. He mistakenly interpreted the emission as high-energy gamma rays. In 1932, Irène Joliot-Curie (1897–1956) and Frédéric Joliot-Curie (1900–58) repeated Bothe's experiment using a more intense polonium source of alpha particles and adding a paraffin (wax) target behind the beryllium metal target. They observed protons emerging from the paraffin but then incorrectly assumed Bothe's postulated high-energy gamma rays were simply causing some type of Compton effect in the wax and knocking protons out of the paraffin. Madam Curie's older daughter and her husband were capable scientists who would win the 1935 Nobel Prize in chemistry "for synthesis of new radioactive elements." They just hastily reached the wrong conclusion about the interesting results of this particular experiment. The misinterpretations of experimental results by both Bothe and the Joliot-Curies set the stage for Chadwick to discover the neutron.

In 1932, Chadwick received news about the recent French experiment. He did not agree with their conclusions, so he immediately repeated the same experiment at the Cavendish Laboratory, but with the specific goal of looking for Rutherford's long-postulated neutral particle, the neutron. His experiments were successful, and he was able not only to prove the existence of the neutron but also that this neutral particle was slightly more massive than the proton. On February 27, 1932, Chadwick announced his important finding in the paper "Possible Existence of a Neutron."

He received the 1935 Nobel Prize in physics for the discovery. His great achievement marks the beginning of modern nuclear science because the neutron allowed scientists to more effectively probe the atomic nucleus and unlock its secrets.

Chadwick departed the Cavendish Laboratory in 1935 to accept the chair of physics at the University of Liverpool. There, he constructed the first cyclotron in the United Kingdom and supported the preliminary study by the Austrian-British physicist Otto Frisch (1904–79) and the German-British physicist Rudolph Peirls (1907–55) concerning the feasibility of making an atomic bomb. Since energy for this type of weapon of mass destruction actually comes from the atomic nucleus, the term *nuclear weapon* is technically more accurate. However, both terms have experienced popular, almost interchangeable, use since World War II. Chadwick spent a good portion of World War II in the United States as the head of the British mission that supported the Manhattan Project (discussed below).

Chadwick's work allowed physicists to complete the basic model of the nuclear atom. His discovery of the neutron also set in motion a wave of neutron-related research in the 1930s and 1940s by brilliant scientists such as the Italian-American physicist Enrico Fermi (1901–54). The neutron resulted in the discovery of nuclear fission by Otto Hahn in 1938 and in the operation of the world's first nuclear reactor by Fermi in 1942.

SEGRÈ AND THE FIRST ARTIFICIAL ELEMENT

As the dark clouds of war began to gather over Europe in the late 1930s, the international scientific community busily focused on trying to produce new elements by bombarding target materials (such as uranium) with neutrons and other energetic nuclear particles. The Italian-American physicist Emilio Segrè produced the first new element by artificial means.

In 1937, he discovered element 43, which he named technetium, from the Greek word meaning "artificial." Earlier, while visiting the University of California, Berkeley, he had obtained an accelerator-exposed piece of scrap metal consisting of molybdenum (Mo) foil that had been used as a particle beam deflector in the facility's cyclotron. (Molybdenum is element 42.) Segrè took the neutron-bombarded sample of molybdenum back to his laboratory in Italy, where he and fellow scientist Carlo Perrier (1886–1948) performed a successful search for the small quantity of human-manufactured technetium embedded within the neutron-irradiated molybdenum.

Because of its radioactive nature, technetium (Tc) does not exist naturally on Earth in any appreciable quantities. Technetium is the lowest atomic number chemical element that has no stable isotope. With a half-live of 4.2 million years, technetium-98 is the most enduring radioisotope of the unstable element. Technetium-99m is a short-lived (about six-hour half-life) gamma ray–emitting, metastable nuclear isomer that plays a major role in nuclear medicine. While experiencing an internal transition (IT), technetium-99m emits a gamma ray and then transforms into the radioisotope technetium-99. Technetium-99 is a weak beta particle emitter with a half-life of approximately 211,000 years. In the 1930s and 1940s, technetium was a very scarce artificial element, but with the advent of commercial nuclear reactors, technetium became available in significant quantities as part of the fission product inventory contained in spent nuclear reactor fuel.

During another visit to the United States in 1938, Benito Mussolini's fascist government decided to remove Segrè from his academic position at the University of Palermo. Undaunted, the gifted nuclear physicist simply stayed in the United States and made many contributions to science—including the isolation and identification of technetium-99m, the most widely used medical radioisotope. When the United States entered World War II, Segrè participated as a nuclear scientist in the American atomic bomb program.

THE DISCOVERY OF NUCLEAR FISSION

Just before Christmas in 1938, a very interesting but also very disturbing scientific message trickled out of Nazi Germany. The significance of this message traveled to the United States by way of the informal network of scientists who were greatly concerned about nuclear physics developments inside Nazi Germany.

The German radiochemists Otto Hahn and Fritz Strassmann (1902–80) made an unexpected discovery at the Kaiser Wilhelm Institute of Chemistry in Berlin. While they were bombarding various elements with neutrons, they noticed that the majority of neutron-bombarded elements changed only somewhat during their experiments. However, uranium nuclei changed significantly—almost as if the uranium nuclei were split-

NUCLEAR FISSION

During the process of nuclear fission, the nucleus of a heavy element, such as uranium (U) or plutonium (Pu), is bombarded by a neutron, which it then absorbs. The resulting compound nucleus is unstable and soon breaks apart (or fissions), forming two lighter nuclei (called *fission products*) and releasing additional neutrons.

In a properly designed nuclear reactor, the fission neutrons are used to sustain the fission process in a controlled chain reaction. The fission process is also accompanied by the release of a large amount of energy, typically 200 million electron volts (MeV) per reaction. Much of this energy appears as the kinetic (or motion) energy of the fission product nuclei, which kinetic energy is then converted to

A Hanford Site worker safely handles a two-kilogram *button* of plutonium metal. *(Department of Energy)*

thermal energy as the fission products slow down in the reactor fuel material. Engineers use a circulating coolant to remove thermal energy from the reactor's core and then apply the heated coolant in the generation of electricity or in industrial processes requiring heat.

Energy is released during the nuclear fission process because the total mass of the fission products and neutrons after the reaction is less than the total mass of the original neutron and heavy fissionable nucleus that absorbed it. Einstein's mass-energy equivalence formula determines the specific amount of

(continues)

(continued)

energy released in a fission reaction. Nuclear fission can occur spontaneously in heavy elements but is usually caused when nuclei absorb neutrons. In some (rarer) circumstances, nuclear fission may also be induced by very energetic gamma rays during a process scientists call *photofission.*

The most important fissionable (or fissile) materials are the radioisotopes uranium-235 and plutonium-239. Uranium-235 occurs as a very small percentage (about 7 atoms in 1,000) of all uranium atoms found in nature. The bulk of the natural uranium atoms are the radioisotope uranium-238. Uranium-238 atoms are not fissionable under most conditions found in commercial nuclear reactors. However, scientists call uranium-238 a *fertile nuclide,* meaning that when it absorbs a neutron, instead of undergoing fission, it eventually transforms (by way of the formation and decay of neptunium-239) into another fissile nuclide, plutonium-239. Metric tons of plutonium-239 now exist because of the neutron capture and elemental transformation process that takes place in the core of nuclear reactors.

ting into two more or less equal halves. Hahn was greatly disturbed by the results, especially when he tried to explain the quantity of barium that kept appearing in the uranium target material, but Hahn could not bring himself to accept the fantastic possibility that neutron bombardment was causing the uranium nucleus to split. In an attempt to resolve the puzzle, Hahn sent a letter describing the recent experiments and the unusual results to his former long-time laboratory coworker Lise Meitner (1878–1968).

Meitner, an Austrian Jew raised as a Protestant, had hastily abandoned her research activities with Hahn in 1938 and fled for her life to Stockholm, Sweden. She had no choice. When Nazi Germany annexed Austria, the tenuous safety afforded by her Austrian citizenship immediately vanished. Meitner read Hahn's letter and suddenly saw the answer, as incredible as it might be. To confirm her suspicions, she visited her nephew Otto Frisch, who was working at the time with Niels Bohr in Copenhagen, Denmark. They discussed the Hahn-Strassmann experiment and mutually concluded that the uranium nucleus was indeed splitting, or undergoing fission. Science historians attribute this particular use of the term *fission* to the American biologist William Archibald

Arnold (1904–2001), who happened to be in Copenhagen collaborating with George de Hevesy. Arnold suggested the term because of its parallel use in biology to describe the splitting in two of living cells. The physicists liked Arnold's analogy, and the term *fission* entered the lexicon of nuclear physics.

Using Bohr's liquid drop model of the atomic nucleus and referencing the unexplained but persistent presence of barium in the Hahn-Strassmann uranium bombardment experiments, Meitner and Frisch publicly proposed the possibility of nuclear fission. Their ideas appeared in a landmark paper published in the February 11, 1939, issue of *Nature*. They also postulated that about 200 million electron volts (MeV) of energy should be released each time a uranium nucleus split into approximately equal halves.

DAWN OF THE NUCLEAR AGE

Throughout history, only a few events have dramatically altered the course of human history. The discovery and use of fire in prehistoric times is one example. The Bronze Age and the Iron Age represent two other examples. This section summarizes the two defining moments that brought about the dawn of the nuclear age in the 20th century. Amazingly, the events occurred less than three years apart during World War II. Scientists accomplished the technical milestones as part of the secret American atomic bomb project, code-named the Manhattan Project.

In late 1942, Enrico Fermi led a small team of scientists at the University of Chicago in operating the world's first nuclear reactor. The first human-initiated, self-sustaining nuclear chain reaction occurred under strict wartime secrecy in very unassuming surroundings. Construction of Chicago Pile One (CP-1) had started on November 16, 1942, under Fermi's supervision on the campus of the University of Chicago. In the squash court under the stands of the Stagg Field stadium, Fermi's staff began placing graphite bricks in a circular layer to form the base of the atomic pile. When completed, the carefully assembled pile contained 57 layers of uranium metal and uranium oxide embedded in graphite blocks. A wooden structure supported the towering graphite pile, which contained 22,000 uranium slugs and required 380 tons (344.8 tonnes) of graphite, 40 tons (36.3 tonnes) of uranium oxide, and six tons of uranium metal.

All was ready on the bitterly cold morning of December 2, 1942, for the start of the experiment that changed the course of human history. At

3:20 P.M. (local time), Fermi instructed an assistant to move the control rod out another 5.9 inches (15 cm). The neutron count monitoring instruments responded but eventually leveled off. Fermi waited for five minutes and then asked that the control rod be taken out another foot (0.3 m). As the rod was being withdrawn, Fermi turned to Arthur Holly Compton, the project director, and told him that the reactor would now become self-sustaining, as indicated by the trace of the graph climbing and continuing to climb with no evidence of leveling off.

Paying particular attention to the rate of rise of the neutron counts per minute, Fermi used his slide rule to make some last-minute calculations. Suddenly, Fermi closed his slide rule, glanced at the instruments, and quietly announced that the reaction was self-sustaining. The small group of scientists involved in the experiment continued to watch in solemn awe as the first human-made nuclear reactor operated for 28 minutes. Then, at 3:53 P.M., Fermi ordered the chain reaction stopped. On Fermi's order, another scientist dutifully inserted the safety rod.

Fermi and his team were the first people to successfully initiate a self-sustaining fission chain reaction and release the energy deep within the atomic nucleus in a controlled manner. Science historians generally regard the date as the dawn of the nuclear age. On that historic day, CP-1 operated at the very modest power level of just half a watt (0.5 watt). Ten days later, Fermi's team operated CP-1 at 200 watts and could have increased the power level further, but radiation safety issues suggested a more prudent course of action.

Fermi's reactor demonstrated the key technology needed for the production of large quantities of plutonium. Manhattan Project physicists highly prized this human-made transuranium element (discovered in early 1941) as a more efficient candidate nuclear fuel for the American atomic bomb. Fermi's success encouraged the military director of the Manhattan Project, General Leslie Groves (1896–1970), to start construction of the project's large-scale plutonium production reactors in Hanford, Washington State. Plutonium produced at Hanford would soon fuel the world's first nuclear detonation.

At 5:30 A.M. (local time) on July 16, 1945, the age of nuclear warfare began. While military and scientific members of the Manhattan Project anxiously watched, a plutonium-fueled implosion test device exploded over the southern New Mexican desert. The blast vaporized the tall support tower and turned the asphalt around its base into fused green-hued

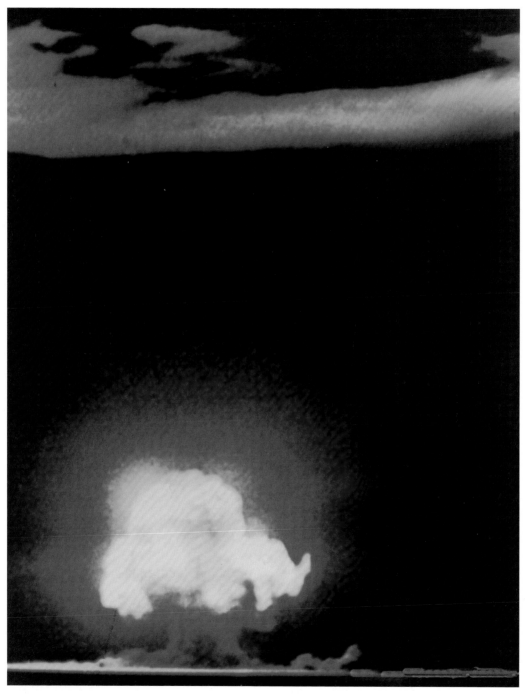

Dawn of the nuclear weapons era—the Trinity explosion on July 16, 1945 *(DOE/LANL)*

sand, later dubbed trinitite. The world's first nuclear explosion, code-named Trinity, released the energy equivalent of about 18.6 kilotons (kT) of the chemical high explosive trinitrotoluene (TNT).

In the early dawn, the Trinity fireball proved brighter than many suns, making some observers suffer temporary blindness despite their darkened protective glasses. Seconds after the explosion, a huge blast wave came roaring across the desert floor and tumbled some unprepared observers to the ground. All felt the searing heat radiating from the giant orange and yellow fireball. Seconds later, the fireball rose and transformed into a mushroom-shaped cloud, imprinting indelibly on human consciousness the universally recognized nuclear age symbol for large-scale death and destruction.

Throughout the Manhattan Project, fascinating new scientific discoveries occurred at a rate faster than scientists or engineers could comfortably absorb. Under the goal-oriented leadership of General Groves, physicists and chemists were given no time to wander down intellectual side roads or to confirm fundamental concepts through the lengthy, conservative process of hypothesis formation, laboratory experimentation, technical publication, and peer review. In 1939, Einstein sent his famous letter to President Franklin D. Roosevelt encouraging the development of an American atomic bomb before Nazi Germany. At that time, no one realized the enormous engineering difficulties involved in taking the basic concepts of fission and of a chain reaction and translating these from the frontiers of nuclear physics into a tangible, aircraft deliverable device that could release an enormous amount of energy in an explosively rapid yet predictable fashion.

The Trinity blast heralded the dawn of a new age in warfare—the age of nuclear weaponry. From that fateful moment on, human beings were capable of unleashing wholesale destruction in an instantaneous level of violence unavailable in all earlier periods of history. While observing the Trinity fireball, the American physicist J. Robert Oppenheimer (1904–67), who led the team of atomic bomb scientists working at Los Alamos, New Mexico, recalled the ancient Hindu declaration, "I am become Death, the destroyer of worlds."

In Greek mythology, the Titan Prometheus stole fire from Mount Olympus, home of Zeus and the other Olympian gods, and bestowed it as a gift upon humankind. As the legend goes, Zeus became angry and severely punished Prometheus, but the gift of fire gave human beings the advantage they needed to survive and progress. Science historians some-times refer to the nuclear physicists who first unleashed the energy hidden within the atomic nucleus as the new Prometheans—people who have given the human race the gift of a new type of fire.

CHAPTER 9

Contemporary View of Matter

This chapter explores how the contemporary understanding of matter on the smallest (or quantum) scale emerged in the 20th century. Especially remarkable progress took place in the decades immediately following World War II. The chapter also describes some of the unusual aspects and consequences of the intriguing world of quantum mechanics.

INTRIGUING WORLD OF QUANTUM MECHANICS

The world of quantum mechanics is quite alien to daily human experiences. A person's senses generally suggest that large-sized (macroscopic) objects consist of continuous material. There does not appear to be a specific advantage in thinking about atoms and their individual behaviors during normal daily activities. When holding an aluminum baseball bat, a person does not usually think of the myriad of atoms with their clouds of electrons and tiny (but dense) atomic nuclei that actually make up the object. The person just wants to swing the bat and hit the incoming ball.

Scientists now recognize that all matter is quantized. This means that at the atomic scale, matter comes in extremely small bundles characterized by electrons whirling in clouds around extremely small, dense central cores of matter called nuclei. Electric charge is quantized. Each electron possesses a fundamental negative charge of one (–1), and each proton possesses a fundamental positive charge of one (+1). At the atomic scale, energy is also quantized. According to quantum theory, as electrons orbit the nucleus of an atom, they can reside only in certain allowed energy states. As electrons transition (or quantum jump) from higher to

TAKING A REALLY CLOSE-UP LOOK AT MATTER

Modern physics suggests this contemporary view of matter. Consider a very tiny drop of dew (about 3.28×10^{-3} feet [1 mm] in diameter) hanging on a plant. Scientists say that this dew drop contains an incredibly large number of water molecules, each made up of two hydrogen atoms and one oxygen atom. Every hydrogen atom (about 3.28×10^{-10} feet [1×10^{-7} mm] in diameter) consists of an orbiting electron with a proton as its nucleus. The proton, in turn, has a diameter of about 3.28×10^{-15} feet (1×10^{-12} mm) and has its own collection of subnuclear particles, consisting of a combination of three quarks held together by gluons. Finally, the "up quark" depicted here is an incredibly tiny piece of matter, having a diameter of less than 3.28×10^{-18} feet (1×10^{-15} mm).

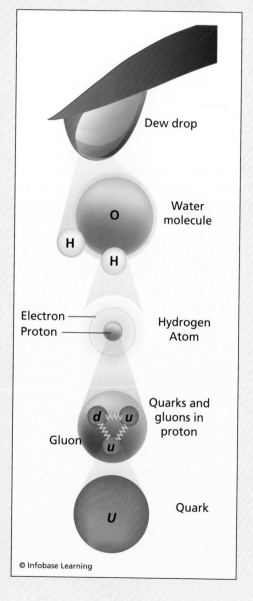

Looking really closely at a very tiny drop of dew (DOE/FNAL modified by author)

© Infobase Learning

lower energy states, they emit discrete packets of electromagnetic energy called photons.

Supporting the overall surge in understanding were experimental physicists such as Sir John Cockcroft, Ernest Walton, and Ernest O. Lawrence (1901–58) who developed pioneering machines and instruments that made the exciting world of subatomic physics available to other scientists. Theoretical physicists such as Murray Gell-Mann (1929–   ) examined the avalanche of data concerning nuclear particles—sometimes termed the *nuclear particle zoo*—and then suggested that supposedly elementary particles such as neutrons and protons actually had structure and smaller particles (called *quarks*) within them. The sum of these activities supported the maturation of quantum mechanics and resulted in the emergence of the standard model of fundamental particles.

While the often bizarre behavior of matter at the atomic and subatomic levels defies human intuition and daily experience, quantum mechanics nevertheless greatly influences the lives of almost every person on Earth.

PHOTOELECTRIC EFFECT

When sufficiently energetic photons fall upon the surface of certain substances, especially shiny metals, orbiting electrons can completely absorb the incident photons and then depart their parent atoms in the target material. Scientists call this phenomenon the *photoelectric effect*. It plays an important role in many modern electronic devices.

While experimenting with ultraviolet radiation in 1887, Heinrich Hertz observed that the incident ultraviolet light was releasing electric charges (later called electrons) from the surface of a thin metal target. Unfortunately, Hertz did not recognize the significance of this phenomenon, nor did he pursue further investigation of the photoelectric effect.

In 1905, Albert Einstein wrote a famous paper about the photoelectric effect, linking the phenomenon with Max Planck's idea of quantum packets of energy called photons. Einstein received the 1921 Nobel Prize in physics for his explanation of the photoelectric effect. His explanation involved the conservation of energy principle and the postulation of single photon absorption by atoms in the target material. By successfully explaining this important phenomenon, Einstein provided credibility to the emerging scientific field of quantum mechanics.

The global information revolution depends on how devices such as transistors manipulate individual electric charges on the atomic scale. When people enjoy using their digital cameras, they are taking advantage of something called the photoelectric effect. While difficult to perceive from ordinary experience, it is truly a quantum mechanical world.

RISE OF QUANTUM MECHANICS

Quantum mechanics and atomic theory matured in the 1920s through brilliant contributions from scientists such as the French physicist Prince Louis-Victor de Broglie (1892–1987), the German theoretical physicist Werner Heisenberg, the Austrian physicist Erwin Schrödinger (1887–1961), and the British physicist Paul Dirac (1902–84).

Louis-Victor de Broglie earned his doctorate in physics at the Sorbonne (Paris, France) in 1924. Einstein's description of the photoelectric effect stimulated de Broglie's theoretical work and led him to suggest that particles can behave like waves and waves like particles. He postulated that an electron could exhibit wavelike properties. He further suggested that the wavelength associated with an electron ($\lambda_{electron}$) is given by Planck's constant (h) divided by the electron's momentum (p). In 1927, other physicists, including Nobel laureate Sir George Thomson (1892–1975), experimentally observed the wavelike behavior of electrons. De Broglie's work served as the intellectual stimulus for the development of quantum wave mechanics, a theory that essentially transformed scientific understanding of physical phenomena at the atomic scale. He received the 1929 Nobel Prize in physics for his discovery of the wave nature of electrons.

Werner Heisenberg received his doctoral degree in physics in 1923 but was almost disqualified for not completing all his laboratory work. In 1925, he began refining the Bohr-Rutherford atom model and developed (along with other German physicists) a form of quantum mechanics known as matrix mechanics. Two years later, Heisenberg used theory to further explore the quantum mechanical behavior of electrons. This effort resulted in his famous uncertainty principle, which eventually became accepted as a fundamental physical law.

Heisenberg's uncertainty principle postulates that at the atomic (quantum) scale, any simultaneous measurement of a particle's properties (such as position [x] and momentum [p]) basically alters the properties being measured. Heisenberg stated that measured values cannot be simultaneously assigned to the position (x) and the momentum (p) of a particle with unlimited precision. Scientists often express this statement as the uncertainty of position (Δx) times the uncertainty in momentum (Δp) $\geq h/2\pi$. Note the significance of Planck's constant (h) in the uncertainty principle of quantum mechanics. Heisenberg received the 1932 Nobel Prize in physics for his pioneering theoretical work in quantum mechanics.

In 1926, Erwin Schrödinger published a series of scientific papers that served as the foundation of quantum wave mechanics. His work was partially inspired by de Broglie's hypothesis of the wavelike behavior of particles under certain (quantum-level) circumstances. Schrödinger's famous wave equation correctly described the energy levels of the electrons orbiting the nucleus of the hydrogen atom without using the quantum jump assumption proposed by Niels Bohr.

Reflecting some of the intellectual fermentation of the times, Schrödinger preferred his deterministic approach to quantum mechanics rather than the more mathematical and statistical approach being championed by Heisenberg and other physicists. Despite his continued misgivings about the statistical interpretation of quantum mechanics, Schrödinger shared the 1933 Nobel Prize in physics with Paul Dirac for the development of important new forms of the atomic theory of matter. Schrödinger's quantum wave mechanics provided a new system for studying the properties of atoms and molecules under various external conditions.

The wave mechanics versus matrix mechanics approach to quantum theory created a great intellectual schism in physics in the late 1920s. However, the crisis eventually abated when physicists realized that although Heisenberg's matrix mechanics approach and Schrödinger's wave mechanics approach had different starting points and emerged as a result of different thought processes, the two techniques actually produced the same results when they investigated the same quantum physics problem.

Introducing special relativity into the quantum world, Paul Dirac developed a form of relativistic quantum mechanics in the late 1920s. Dirac's work produced solutions that suggested the existence of antimatter states. In 1930, he published *The Principles of Quantum Mechanics*. That same year, he also suggested that photons of sufficient energy might transform into an electron-positron pair, and vice versa. In 1932, while investigating cosmic-ray interactions with the atmosphere using cloud chamber detectors, the American physicist Carl David Anderson (1905–91) discovered the first antimatter particle—the positively charged electron, or *positron*. Dirac's theoretical speculations about the existence of antimatter particles proved correct, and he coshared the 1933 Nobel Prize in physics with Schrödinger.

As a result of Dirac's theoretical work and Anderson's discovery of the positron, scientists gradually began to recognize that every particle has a corresponding antiparticle. For charged particles, the companion antiparticle has the same mass and spin but oppositely signed electric charge. Uncharged particles (such as the neutron) have an antimatter companion that has the same mass but opposite signs on its quantum numbers. When a particle meets its antiparticle, they annihilate each other, and their combined energies appear in the form of pure energy. Neglecting any particle (or antiparticle) kinetic energy for the moment, when an electron and positron annihilate each other, the matter-antimatter pair is replaced by a pair of 0.51-MeV gamma rays that depart the annihilation event in opposite directions. Two energetic (0.51 MeV) gamma rays appear because that amount of radiant energy corresponds to the rest mass energy equivalents of the electron and the positron.

Anderson shared the 1936 Nobel Prize in physics with the Austrian-American physicist Victor Francis Hess (1883–1964). Anderson received the award for his discovery of the positron, while Hess received the award for his discovery of cosmic radiation. Hess's discovery of cosmic rays, using numerous daring balloon flights between 1912 and 1913, turned Earth's atmosphere into a giant natural laboratory for particle physicists. Before the development of high-energy particle accelerators, the study of cosmic ray interactions with atoms of the atmosphere opened the door to many new discoveries about the atomic nature of matter.

ACCELERATORS

As quantum mechanics matured, other researchers began to construct machines called particle accelerators that allowed them to hurl high-energy subatomic particles at target nuclei in an organized and somewhat controllable attempt to unlock additional information about the atomic nucleus.

The particle accelerator is one of the most important tools in nuclear science. Before its invention in 1932, the only known, controllable sources of particles that could induce nuclear reactions were the natural radioisotopes that emitted energetic alpha particles. Nuclear scientists also tried to gather fleeting peeks at energetic nuclear reactions by examining the avalanche of short-lived nuclear particles that rippled down through Earth's atmosphere when extremely energetic cosmic rays smashed into Earth's upper atmosphere.

Modern accelerators have particle energies and beam intensities powerful enough to let physicists probe deep into the atomic nucleus. By literally "smashing atoms," physicists can now examine the minuscule

LAWRENCE'S AMAZING LABORATORIES

Ernest Orlando Lawrence sponsored an enormous wave of research and discovery in nuclear physics when he invented the cyclotron, a new type of high-energy particle accelerator. In 1928, Lawrence accepted an appointment as associate professor of physics at the University of California, Berkeley. Two years after his initial academic appointment, he advanced to the rank of professor, becoming the youngest full professor on the Berkeley campus. In 1936, he also became director of the university's Radiation Laboratory and remained in both positions until his death.

Lawrence's early nuclear physics research involved ionization phenomena and the accurate measurement of the ionization potentials of metal vapors. In 1929, he invented the cyclotron, an important research device that uses magnetic fields to accelerate charged particles to very high velocities without the need for very high voltage supplies. Developed and improved in collaboration with other researchers, Lawrence's cyclotron became the main research tool at the Radiation Laboratory and helped revolutionize nuclear physics for the next two decades. As more powerful cyclotrons appeared, scientists from around the world came to work at Lawrence's laboratory and organized themselves into highly productive research teams. Not only did Lawrence's cyclotron provide a unique opportunity to explore the atomic nucleus with high-energy particles, his laboratory's research teams also kept producing important results that were beyond the capabilities of any single individual working alone.

Lawrence received the 1939 Nobel Prize in physics for the invention of the cyclotron and the important results obtained with it. Because of wartime travel hazards, Lawrence could not go to Stockholm for the presentation ceremony. Just prior to direct American involvement in World War II, the American scientists Edwin McMillan (1907–91) and Glenn Seaborg (1912–99) discovered the structure that lies within an individual proton or neutron. The resultant soup of quarks and gluons resembles conditions at the birth of the universe.

Common to all accelerators is the use of electric fields for the acceleration of charged particles. The manner in which different machines use electric fields to accelerate particles varies considerably. This section

first two transuranic elements (neptunium and plutonium) at the Radiation Laboratory. The discovery of plutonium-239 and its superior characteristics as a fissile material played a major role in the development of the American atomic bomb. Lawrence's Radiation Laboratory became a secure defense plant during World War II, and its staff members made many contributions to the overall success of the Manhattan Project.

After the war, Lawrence returned to the business of making particle accelerators and supported the concept of the synchrocyclotron. The new machine was a way of getting to even higher particle energies. A vastly improved bubble chamber developed by Luis Alvarez (1911–88) at the Radiation Laboratory opened up the nuclear particle zoo.

For political and technical reasons, Lawrence vigorously supported development of the American hydrogen bomb and the establishment of a second major nuclear weapons design laboratory in Livermore, California. Responding to Lawrence's recommendations, the United States Atomic Energy Commission (USAEC) created this second nuclear weapons complex to promote thermonuclear weapons development in the 1950s. A colleague and friend of Lawrence, the Hungarian-American physicist Edward Teller (1908–2003), served as the new laboratory's director from 1958 to 1960.

While at an important international conference on nuclear arms control held in Geneva, Switzerland, in summer 1958, Lawrence fell gravely ill and was rushed back to the United States for medical treatment. Following surgery, he died on August 27, 1958, in Palo Alto, California. Today, two major national laboratories, the Lawrence Berkeley National Laboratory (LBNL) and the Lawrence Livermore National Laboratory (LLNL), commemorate his contributions to nuclear technology. In his honor, scientists named element 103 lawrencium (Lr).

briefly discusses the basic principles behind the operation of some of the most familiar accelerators used in nuclear science.

The most straightforward type of accelerator is the Cockcroft-Walton machine that appeared in 1932 and opened up a new era in nuclear research. The basic device applied a potential difference between terminals. To obtain an accelerating voltage difference of more than about 200 kilovolts (kV), scientists employed one or more stages of voltage doubling circuits. Sir John Douglas Cockcroft and Ernest Thomas Walton used their early machine to perform the first nuclear transmutation experiments with artificially accelerated nuclear particles (protons). For their pioneering research involving the transmutation of atomic nuclei by artificially accelerated atomic particles, they shared the Nobel Prize in physics in 1951. The Cockcroft-Walton (direct current) accelerator is still widely used in research.

The radio frequency (RF) linear accelerator repeatedly accelerates ions through relatively small potential differences, thereby avoiding problems encountered with other accelerator designs. In a linear accelerator (or *linac*), an ion is injected into an accelerating tube containing a number of electrodes. An oscillator applies a high-frequency alternating voltage between groups of electrodes. As a result, an ion traveling down the tube will be accelerated in the gap between the electrodes if the voltage is in phase. In a linear accelerator the distance between electrodes increases along the length of the tube so that the particle being accelerated stays in phase with the voltage.

The availability of high-power microwave oscillators after World War II allowed relatively small linear accelerators to accelerate particles to relatively high energies. Today, there are a variety of large linacs, both for electron and proton acceleration as well as several heavy-ion linacs. The Stanford Linear Accelerator (SLAC) at Stanford University in California is a 1.86-mile (3-km)-long electron linac. This machine can accelerate electrons and positrons to energies of 50 GeV.

The cyclotron, invented by the American physicist Ernest Orlando Lawrence in 1929, is the best known and one of the most successful devices for the acceleration of ions to energies of millions of electron volts. The cyclotron, like the RF linear accelerator, achieves multiple acceleration of an ion by means of a radio frequency–generated electrical field, but in a cyclotron, a magnetic field constrains the particles to move in a spiral path. Ions are injected at the center of the magnet between two circular electrodes (called *Dees*). As a charged particle spirals outward, it gets accelerated each time it crosses the gap between the Dees. The time it takes a particle to complete an orbit is constant, since the distance it travels increases at the same rate as its velocity, allowing it to stay in phase with the radio frequency signal. The favorable particle acceleration conditions break down in the cyclotron when the charged particle or ion being accelerated reaches relativistic energies. Despite this limitation, cyclotrons remain in use throughout the world supporting nuclear research, producing radioisotopes, and accommodating medical therapy.

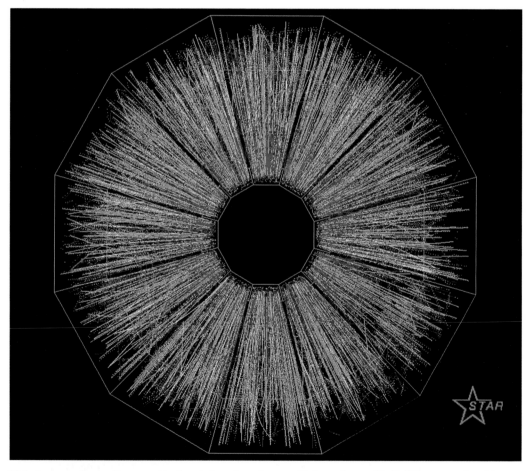

This image shows the result of a violent head-on collision of two 30-GeV beams of gold atom nuclei in the Relativistic Heavy Ion Collider (RHIC) at Brookhaven National Laboratory in New York. For a fleeting moment during this energetic collision, scientists created a gluon-quark plasma. *(DOE/BNL)*

Scientists developed the synchroton to overcome the energy limitations that special relativity imposed on cyclotrons. In a synchrocyclotron, the radius of a particle's orbit is kept constant by a magnetic field that increases with time as the momentum of the particle increases. An RF oscillator supports particle acceleration by supplying an energy increment each time the particle crosses an accelerating gap. Nuclear scientists using the Relativistic Heavy Ion Collider (RHIC) at Brookhaven National Laboratory in New York routinely collide two beams of ions ranging from protons to gold with energies up to 100 GeV. Researchers use such energetic collisions in their ongoing attempts to create temperatures and densities sufficiently high to reach (at least for a very brief moment) the quark-gluon plasma phase of nuclear matter.

The Thomas Jefferson National Accelerator Facility in Newport News, Virginia, is the most recent nuclear accelerator to become operational in the United States. At this accelerator, an electron beam travels through several linear accelerators (linacs). The accelerator employs superconducting, radio frequency technology to drive electrons to higher and higher energies with a minimum input of electrical power. An important feature of this accelerator is the fact that the machine produces a continuous electron beam. This ensures that each electron interaction with a nucleus can be separated sufficiently in time that scientists can measure the entire reaction.

Founded in 1954, the European Organization for Nuclear Research (CERN, a French-language acronym) is the world's largest particle physics research center. Located just outside Geneva, Switzerland, the large accelerators of the international research center actually straddle the Franco-Swiss border. CERN's Large Hadron Collider (LHC) is a gigantic particle accelerator in which two beams of subatomic particles (called *hadrons*) travel in opposite directions inside the circular accelerator, gaining energy with every lap. By colliding two beams head-on at very high energies, researchers can recreate the gluon-quark plasma conditions that existed just after the big bang. A large number of special detectors collect information from these energetic conditions and enable scientists to improve their models of matter at the most minuscule scale deep within the nuclear structure of atoms.

The basic mission of the Fermi National Accelerator Laboratory (FNAL, or Fermilab) in Batavia, Illinois, is to build and operate accelerators, detectors, and supporting facilities to conduct pioneering research in high-energy physics, research that explores the innermost structure

An inside view of the drift tube in the older 200-MeV section of the linear accelerator (linac) at the Department of Energy's Fermi National Accelerator Laboratory in Batavia, Illinois *(DOE/FNAL)*

of matter. Fermilab's Tevatron is the highest-energy particle accelerator and collider in the United States and second-largest in the world, behind CERN's LHC. At Fermilab and other very large accelerator facilities, linacs are often used as injectors to gigantic circular accelerators. Scientists working at Fermilab have contributed a great deal to the understanding of the basic structure of matter. Their discoveries include the bottom (b) and top (t) quarks as well as the tau neutrino (v_τ).

There are many variations of the basic accelerators just described. The particle accelerator remains a basic tool of nuclear science. With them, experimenters can probe deeper and deeper into the nature of matter and begin to replicate the highly energetic, primordial conditions of the very early universe that existed immediately following the big bang.

THE NUCLEAR PARTICLE ZOO

From the 1930s through the early 1960s, scientists successfully developed the technical foundations for modern nuclear science. These developments included the nuclear reactor, the nuclear weapon, and many other applications of nuclear energy. In the process, scientists and engineers used a nuclear model of the atom that assumed the nucleus contained two basic building block nucleons: protons and neutrons. This simple nuclear atom model (based on three elementary particles: the proton, neutron, and electron) is still very useful in discussions concerning matter at the atomic scale.

However, as cosmic-ray research and accelerator experiments began to reveal an interesting collection of very short–lived subnuclear particles, scientists began to wonder what was really going on within the nucleus. They pondered whether some type of interesting behavior involving neutrons and protons was taking place deep within the nucleus.

The "stampede" of new particles began rather innocently in 1930, when Wolfgang Pauli suggested that a particle, later called the neutrino by Enrico Fermi (who used the Italian word for "little neutral one"), should accompany the beta decay process of a radioactive nucleus. These particles were finally observed during experiments in 1956. Today, physicists regard neutrinos as particles that have almost negligible (if any) mass and travel at or just below the speed of light. Neutrinos are stable members of the lepton family.

Two exciting new particles joined the nuclear zoo in 1932, when Carl Anderson discovered the positron and Chadwick the neutron. Knowledge of the neutron and positron encouraged scientists to discover many other strange particles.

One of the most important early hypotheses about forces within the nucleus took place in the mid-1930s, when the Japanese physicist Hideki Yukawa (1907–81) suggested that nucleons (that is, protons and neutrons) interacted by means of an exchange force (later called the strong force). He proposed that this force involved the exchange of a hypothetical subnuclear particle called the pion. This short-lived subatomic particle was eventually discovered in 1947. The pion is a member of the meson group of particles within the hadron family.

Since Yukawa's pioneering theoretical work, remarkable advances in accelerator technology have allowed nuclear scientists to discover several hundred additional particles. Virtually all of these elementary (subatomic) particles are unstable and decay with lifetimes between 10^{-6} and 10^{-23} s. Overwhelmed by this rapidly growing population of particles, nuclear scientists no longer felt comfortable treating the proton and the neutron as elementary particles.

While a detailed discussion of all these exciting advances in the physics of matter on the nuclear scale is beyond the scope of this book, the following discussion should make any "visit" to the nuclear particle zoo more comfortable. First, physicists divide the group of known elementary particles into three families: the photons, the leptons, and the hadrons. The basic discriminating factor involves the nature of the force by which a particular type of particle interacts with other particles. Physicists currently recognize four fundamental forces in nature: gravitation, electromagnetism, the weak force, and the strong force.

Scientists regard the photon as a stable, zero rest mass particle. Since a photon is its own antiparticle, it represents the only member of the photon family. Photons interact only with charged particles, and such interactions take place via the electromagnetic force. A common example is Compton scattering by X-rays or gamma rays. Scientists consider the photon quite unique; they have yet to discover another particle that behaves in just this way.

The lepton family of elementary nuclear particles consists of those particles that interact by means of the weak force. Current interpretations of how the weak force works involve refinements in quantum electrodynamics (QED) theory. This theory, initially introduced by the American physicist Richard Feynman (1918–88) and others in the late 1940s, combines Maxwell's electromagnetic theory with quantum mechanics in a way that implies electrically charged particles interact by exchanging a virtual photon. (Responding to the Heisenberg uncertainty principle, a virtual particle exists for only an extremely short period of time.) There is good experimental verification of QED. Leptons can also exert gravitational and (if electrically charged) electromagnetic forces on other particles. Electrons (e), muons (μ), tau particles (τ), neutrinos (ν), and their antiparticles are members of this family. The electron and various types of neutrinos are stable, while other members of the family are very unstable, with lifetimes of a microsecond (10^{-6} s) or less.

The hadron family contains elementary particles that interact by means of the strong force and have a complex internal structure. This family is further divided into two subclasses: mesons (which decay into leptons and photons) and baryons (which decay into protons). Hadrons may also interact by electromagnetic and gravitational forces, but the strong force dominates at short distances of 3.28×10^{-15} ft (1.0×10^{-15} m) or less. The pion (meson), neutron (baryon), and proton (baryon) are members of the hadron family, along with their respective antiparticles. Most hadrons are very short-lived, with the exception of the proton and the neutron. The proton is stable, and the neutron is stable inside the nucleus but unstable outside the nucleus, exhibiting a half-life of about 12 minutes.

In the early 1960s, the American physicist Murray Gell-Mann and others introduced *quark* theory to help describe the behavior of hadrons within the context of the theory of quantum chromodynamics (QCD). Quark theory suggests that hadrons are actually made up of combinations of subnuclear particles called *quarks,* a term that Gell-Mann adapted from a passage in the James Joyce comic fictional work *Finnegans Wake.* Contemporary quark theory suggests the existence of six types of quarks called up (u), down (d), strange (s), charmed (c), top (t), and bottom (b) as well as their corresponding antiquarks. Scientists first discovered quarks during experiments at SLAC in the late 1960s and early 1970s. The top quark was the last quark to be experimentally discovered. This event took place in 1995 at Fermilab.

Experiments indicate that the up (u) quark has an energy equivalent mass of 3 MeV, a charge of +2/3 (that of a proton), and a spin of 1/2. The down (d) quark has a mass of 6 MeV, a charge of −1/3, and a spin of 1/2. The charm (c) quark has a mass of 1,300 MeV, a charge of +2/3, and a spin of 1/2. The strange (s) quark has a mass of 100 MeV, a charge of −1/3, and a spin of 1/2. The top (t) quark has a mass of 175,000 MeV, a charge of +2/3, and a spin of 1/2. Finally, the bottom (b) quark has a mass of 4,300 MeV, a charge of –1/3, and a spin of 1/2.

The first family of quarks consists of up and down quarks, which are the quarks that join together to form protons and neutrons. The second family of quarks consists of strange and charm quarks, which exist only at high energies. The third family of quarks consists of top and bottom quarks. These last two quarks exist only under very high–energy conditions, as may occur briefly during accelerator experiments.

In their attempt to explain what is happening inside the atomic nucleus, scientists needed to learn more about how quarks make up protons and neutrons. Physicists now postulate that quarks have not only electromagnetic charge (in the somewhat odd fractional values mentioned above), but also

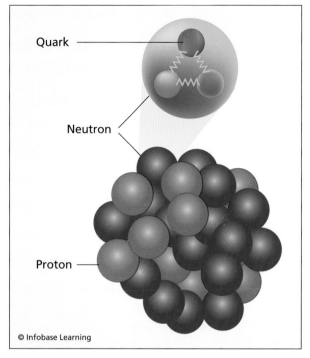

The nucleus of an atom contains many nucleons (protons and neutrons). Nucleons, in turn, are made up of quarks and gluons. *(DOE/BNL)*

an unusual, different type of charge, called *color charge.* It is the force between color-charged quarks that gives rise to the strong force that holds quarks together as they form hadrons. Scientists call the carrier particles associated with the strong force *gluons (g).*

In the standard model of matter (discussed below), scientists assume that quarks and gluons are color-charged particles. Color-charged particles exchange gluons, thereby creating a very strong color force field that binds the quarks together. Note that the term *color charge* as used by physicists does not imply a relationship with the colors of visible light as found in the electromagnetic spectrum. Rather, physicists have arbitrarily assigned various colors to help complete the description of the quantum-level behavior of these very tiny subnuclear particles within hadrons. As a result, quarks are said to have three color charges, red, green, and blue; and antiquarks, three corresponding color charges, antired, antigreen, and antiblue. This additional set of quantum-level properties is an attempt by scientists to bring experimental observations into agreement with Wolfgang Pauli's exclusion principle.

Quarks constantly change their "colors" as they exchange gluons with other quarks. This means that whenever a quark within a hadron absorbs or emits a gluon, that particular quark's color must change in order to conserve color charge and maintain the bound system (that is, protons and neutrons) in a color-neutral state. Scientists also propose that quarks cannot exist individually, because the color force increases as the distance between the quarks increases. The residual strong force between quarks is sufficient to hold atomic nuclei together and overcome any electrostatic repulsion between protons.

THE STANDARD MODEL

Scientists have developed a quantum-level model of matter they refer to as the *standard model*. This comprehensive model explains (reasonably well) what the material world (that is, ordinary matter) consists of and how it holds itself together. As shown in the accompanying figure, physicists need only six quarks and six leptons to explain matter. Despite the hundreds of different particles that have been detected, all known matter particles are actually combinations of quarks and leptons. Furthermore, quarks and leptons interact by exchanging force-carrier particles. The most familiar lepton is the electron (e), and the most familiar force-carrier particle is the photon.

The standard model is a reasonably good theory and has been verified to excellent precision by numerous experiments. All of the elementary particles making up the standard model have been observed through experiments, but the standard model does not explain everything of interest to scientists. One obvious omission is the fact that the standard model does not include gravitation.

In the standard model, the six quarks and the six lepton are divided in pairs, or generations. The lighter and more stable particles compose the first generation, while the less-stable and heavier elementary particles make up the second and third generations. As presently understood by scientists, all stable (ordinary) matter in the universe consists of particles that belong to the first generation. Elementary particles in the second and third generation are heavier and have very short lifetimes, decaying quickly to the next more stable generation.

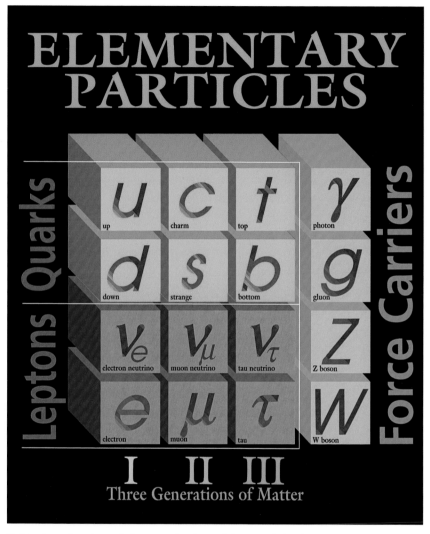

Scientists developed the standard model to explain the complex interplay between elementary particles and force carriers. *(DOE/FNAL)*

The everyday world of normal human experience involves just three of these building blocks: the up quark, the down quark, and the electron. This simple set of particles is all that nature requires to make protons and neutrons and to form atoms and molecules. The electron neutrino (v_e) rounds out the first generation of elementary particles. Scientists have observed the electron neutrino in the decay of other particles. There may be other elementary building blocks of matter to explain dark matter, but scientists have not yet observed such building blocks in their experiments.

Elementary particles transmit forces among one another by exchanging force-carrying particles called bosons. The term *boson* is the general name scientists have given to any particle with a spin of an integral number (that is, 0, 1, 2, etc.) of quantum units of angular momentum. Carrier particles of all interactions are bosons. Mesons are also regarded as bosons. The term honors the Indian physicist Satyendra Nath Bose (1894–1974).

As represented in the accompanying figure, the photon (γ) carries the electromagnetic force and transmits light. The gluon (*g*) mediates the strong force and binds quarks together. The *W* and *Z* bosons represent the weak force and facilitate the decay of heavier (more energetic) particles into lower-mass (less energetic) ones. The only fundamental particle predicted by the standard model that has not yet been observed is a hypothetical particle scientists call the *Higgs boson*. The British theoretical physicist Peter Higgs (1929–   ) hypothesized in 1964 that this type of particle may explain why certain elementary particles (such as quarks and electrons) have mass and other particles (such as photons) do not. If found by research scientists this century, the Higgs boson (sometimes called the *God particle*) could play a major role in refining the standard model and shedding additional light on the nature of matter at the quantum level. If nature does not provide scientists with a Higgs boson, then they will need to postulate other forces and particles to explain the origin of mass and to preserve the interactive components of the standard model, which they have already verified by experiments.

Finally, the force of gravitation is not yet included in the standard model. Some physicists hypothesize that the gravitational force may be associated with another particle, which they call the *graviton*. One of the major challenges facing scientists this century is to develop a quantum formulation of gravitation that encircles and supports the standard model. The harmonious blending of general relativity (which describes gravitation on a cosmic scale) and quantum mechanics (which describes the behavior matter on the atomic scale) would represent another incredible milestone in humankind's search for the meaning of substance.

CHAPTER 10

Manipulating Matter Atom by Atom

This chapter introduces the amazing world of nanotechnology. Scientists define nanotechnology as the manufacture and application of nanoscale machines, electronic devices, and chemical or biological systems—all of which have characteristic dimensions on the order of one to 100 nanometers. One nanometer (nm) is equivalent to a length of 3.281×10^{-9} ft. Another way of expressing this very small dimension is to realize that a typical sheet of paper (including this page) has a thickness of about 100,000 nanometers. There are also 25,400,000 nanometers per inch.

Nanotechnology is a work in progress and represents combined scientific and industrial efforts involving the atomic-level manipulation of matter. As scientists examine the behavior of materials in the Lilliputian realm of nanotechnology, they are discovering unusual chemical, physical, and biological properties. These properties appear to differ in important ways from the way single atoms or molecules behave or the way very large numbers of atoms or molecules behave in bulk to produce the macroscopic features and properties of matter. Breakthroughs in nanotechnology in this century have the potential to significantly improve the understanding and use of materials by future generations of humans.

(continues on page 166)

INTEGRATED CIRCUIT (IC)

An integrated circuit (IC) is an electronic circuit that includes transistors, resistors, capacitors, and their interconnections, all fabricated on a very small piece of semiconductor material (usually referred to as a *chip*). The American electrical engineer Jack S. Kirby (1923–2005) invented the integrated circuit while working on electronic component miniaturization during summer 1958 at Texas Instruments. Although Kirby's initial device was quite crude by today's standards, it proved to be a groundbreaking innovation that paved the way for the truly miniaturized electronics packages that now define the digital information age. Kirby shared the 2000 Nobel Prize in physics "for his part in the invention of the integrated circuit."

As sometimes happens in science and engineering, two individuals independently came up with the same innovative idea at about the same time. In January 1959, another American engineer, Robert N. Noyce (1927–90), while working for a California company named Fairchild Semiconductor, independently duplicated Kirby's feat. In 1971, Noyce, who was then the president and chief executive officer of another California company, Intel, developed the world's first microprocessor. Today, both Kirby and Noyce are recognized as having independently invented the integrated circuit.

The category of an integrated circuit, such as LSI and VLSI, refers to the level of integration and denotes the number of transistors on a chip. Using one common yet arbitrary standard, engineers say a chip has small-scale integration (SSI) if it contains fewer than 10 transistors, medium-scale integration if it contains between 10 and 100 transistors, large-scale integration (LSI) if it contains between 100 and 1,000 transistors, and very large-scale integration (VLSI) if it contains more than 1,000 transistors.

Feynman's futuristic speculations about micromachines stimulated interest in the creation of very tiny devices that could perform useful functions at the near-atomic scale. Within a decade or so, engineers began creating such microsized devices. Today, scientists and engineers construct MEMS

(continues)

(opposite page) The scale of some natural things. The art depicted is about 5 millimeters (mm) in length, while silicon atoms are spaced about 0.078 nanometer (nm) apart. The vertical scale helps differentiate what scientists mean when they say *microscale* and *nanoscale*. (DOE)

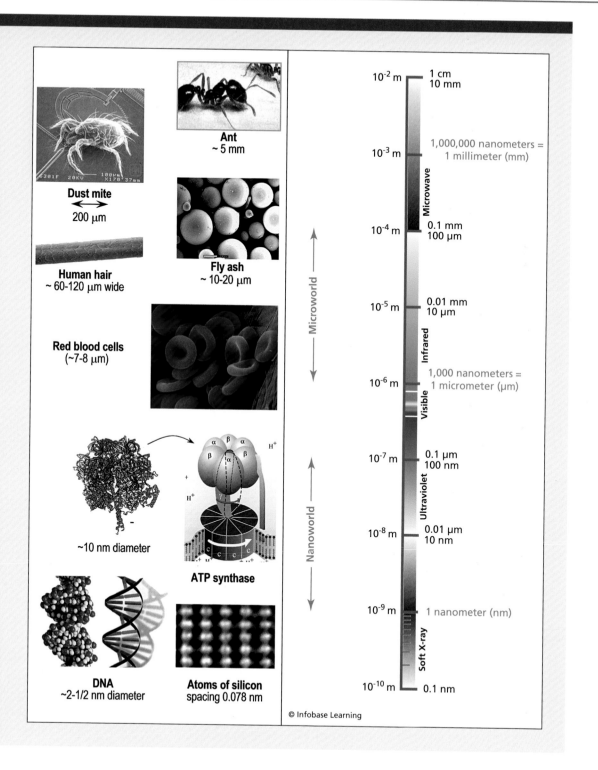

Ant
~ 5 mm

Dust mite
200 μm

Human hair
~ 60-120 μm wide

Fly ash
~ 10-20 μm

Red blood cells
(~7-8 μm)

~10 nm diameter

ATP synthase

DNA
~2-1/2 nm diameter

Atoms of silicon
spacing 0.078 nm

10^{-2} m — 1 cm
10 mm

1,000,000 nanometers =
1 millimeter (mm)

10^{-3} m

Microwave

10^{-4} m — 0.1 mm
100 μm

Microworld

10^{-5} m — 0.01 mm
10 μm

Infrared

1,000 nanometers =
1 micrometer (μm)

10^{-6} m

Visible

10^{-7} m — 0.1 μm
100 nm

Ultraviolet

Nanoworld

10^{-8} m — 0.01 μm
10 nm

10^{-9} m — 1 nanometer (nm)

Soft X-ray

10^{-10} m — 0.1 nm

© Infobase Learning

Head of a pin
1-2 mm

**MicroElectroMechanical
(MEMS) devices**
10 -100 μm wide

Pollen grain
Red blood cells

Zone plate x-ray "lens"
Outer ring spacing ~35 nm

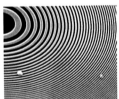

**Self-assembled,
Nature-inspired structure**
Many 10s of nm

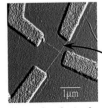

Nanotube electrode

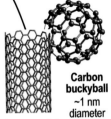

**Carbon
buckyball**
~1 nm
diameter

Carbon nanotube
~1.3 nm diameter

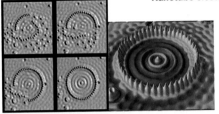

**Quantum corral of 48 iron atoms on copper surface
positioned one at a time with an STM tip**
Corral diameter 14 nm

(continued)

(*microelectromechanical system*) devices with dimensions on the order of 1 to 10 micrometers (ca. 3.281×10^{-6} ft). Science historians generally regard Feynman's lecture as the most convenient milestone marking the beginning of nanotechnology.

According to the National Nanotechnology Initiative (NNI), established by the U.S. government in 2001, nanotechnology is the understanding and control of matter at dimensions between approximately one and 100 nanometers. Since a nanometer (nm) is just one-billionth of a meter (1×10^{-9} m), the term *nanoscale research* refers to science, engineering, and technology performed at the level of atoms and molecules. The practice of nanotechnology involves imaging, measuring, modeling, and manipulating matter within this incredibly tiny physical realm, a world that is certainly well beyond daily human experience. A single gold atom is about one-third of a nanometer in diameter, and 10 hydrogen atoms placed in a row (side by side) would span about one nanometer. By comparison, the width of a single human hair ranges from about 50,000 (blonde) to 150,000 (black) nanometers, and a sheet of paper is about 100,000 nanometers thick. The accompanying figures illustrate the scale of some typical things found in nature and some very small human-made objects.

Scientists are beginning to appreciate that the nanoscale region represents the physical scale at which the fundamental properties of materials and systems are established. How atoms are arranged in nanostructures determines such physical attributes as a material's melting temperature, magnetic properties, and capacity to store charge. Biologists and medical researchers recognize that molecular biology functions, for the most part, at the nanoscale level.

(opposite page) **The scale of some human-made things. Objects depicted range from the head of a pin (about 1 to 2 mm diameter) to a carbon buckyball (about 1 nm diameter).** *(DOE)*

(continued from page 161)

THINKING SMALL

On December 29, 1959, the American physicist and Nobel laureate Richard P. Feynman delivered an inspiring lecture entitled "There's Plenty of Room at the Bottom." He presented this public talk at the California Institute of Technology (Caltech). During his lecture, Feynman suggested that it would be possible (projecting based on late 1950s technology) to write an enormous quantity of information, such as the entire contents of a multivolume encyclopedia, in a tiny space equivalent to the size of the head of a pin. He also speculated about the impact that micro- and nano-sized machines might have. Micro (μ) is the SI unit prefix meaning multiplied by 10^{-6}, and nano (n) means multiplied by 10^{-9}. The word *nano* comes from the ancient Greek word *nanos* (ναvoς), meaning "dwarf."

Feynman suggested that fields such as medicine and information technology (built upon microelectronics) would be revolutionized if these (at the time) hypothetical tiny devices allowed engineers and scientists to manipulate individual atoms and arrange these building blocks of matter into useful devices. As Feynman spoke that December evening, his predicted revolution in microelectronics was actually well underway, sparked by the development of the integrated circuit (IC).

Today's scientists and engineers know a great deal about the physical properties and behavior of matter on the macroscopic scale. They also understand the properties and behavior of individual atoms and molecules when these are considered as isolated, quantum-level systems, but they cannot easily predict the nanoscale behavior of a material by simply extrapolating either its macroscopic (bulk) properties or its quantum-level properties. Rather, at the nanoscale level, collections of atoms exhibit important differences in physical properties and behaviors that cannot be easily explained using traditional theories and models.

Scientists define a *nanoparticle* as very tiny chunk of matter, that is, less than 100 nanometer on a side. This term is less specific than the term *molecule* because it is dimension-specific rather than being related to the chemical composition of the tiny piece of matter. Almost any material substance can be transformed into nanoparticles. Researchers are discovering that nanoparticles of a particular material often behave quite differently than large amounts of the same substance. Some of these differences in physical properties and behaviors result from the continuous modification of a material's characteristics as the sample size changes. Other dif-

NATIONAL NANOTECHNOLOGY INITIATIVE (NNI)

The U.S. government created the National Nanotechnology Initiative (NNI) in 2001 to coordinate federal research and development (R&D) efforts in nanotechnology. The NNI serves as a central focus for communication, cooperation, and collaboration for all federal agencies involved in nanoscale science and engineering. By establishing shared goals, priorities, and strategies, the NNI provides avenues for each individual federal agency (such as the Department of Energy [DOE] and the National Institute of Standards and Technology [NIST]) to leverage the resources of all participating federal agencies.

The NNI has established eight program component areas (PCAs): (1) fundamental nanoscale phenomena and processes; (2) nanomaterials; (3) nanoscale devices and systems; (4) instrumentation research, metrology, and standards; (5) nanomanufacturing; (6) major research facilities and instrument acquisition; (7) environment, health, and safety; and (8) education and social dimensions.

The last PCA is especially significant because it includes research directed at identifying and quantifying the broad implications of nanotechnology for society. This focus includes nanotechnology's social, economic, workforce, educational, ethical, and legal implications. Here, the research challenge facing nanotechnology practitioners is to identify and to quantify, to the greatest extent possible, potential societal developments that could result from discoveries in nanotechnology. With a view toward how technical innovations have influenced societies in the past, NNI participants want to avoid unintended or adverse social impacts.

The NNI currently consists of the individual and cooperative nanotechnology-related activities of 25 federal agencies with a wide range of research and regulatory roles and responsibilities. The NNI's $1.8-billion-dollar budget for 2011 reflects the sustained commitment by the federal government to nanotechnology.

ferences seem to represent phenomena associated with the quantum-level behavior of matter, such as wavelike transport and quantum-size confinement. Scientists have also observed that, beginning at the molecular level, new physical and chemical properties emerge as cooperative interactions, and interfacial phenomena start dominating the behavior of nanoscale molecular structures and complexes.

One of the promises of nanotechnology research is for scientists to develop the ability to understand, design, and control these interesting properties. When this occurs, they will then be able to construct functional molecular assemblies—starting a new revolution in materials science.

By performing nanoscale research, scientists will develop a basic understanding of the physical, chemical, and biological behavior of matter at these very tiny dimensions. Once improved understanding of material behaviors at the nanoscale occurs, scientists and engineers can set about the task of crafting customized new materials and introducing functional atomic-level devices. Researchers project that the potential benefits will influence medicine, electronics, biotechnology, agriculture, transportation, energy production, environmental protection, and many other fields. As part of contemporary nanoscale research activities, scientists and engineers are discovering how molecules organize and assemble themselves into complex nanoscale systems and then begin to function. In the future, they will start constructing a variety of customized quantum devices capable of serving human needs.

NANOMEDICINE

Arab and European alchemists searched in vain for the *panacea*—a special substance that would cure all human illnesses. One very promising line of modern research involves the emerging field of nanomedicine.

Medical researchers start their efforts by considering each human body as a complex system of interacting molecular networks. Then they view a disease (such as cancer) as causing a disruption in one or several of the body's molecular networks. Nanoscale diagnostic devices would allow researchers to study how a particular disease is actually disrupting individual molecular networks. Once better understood, the disease-induced disruption could be prevented or else attacked and neutralized. Specially designed nanoscale devices might also manipulate distressed molecules or deliver minute treatments at precise locations, thereby avoiding damage to healthy and properly functioning portions of a particular molecular network.

VERY SPECIAL MICROSCOPES

At the end of the 16th century, Dutch opticians fabricated the first optical (visible light) microscopes. Up until then, the naked eye could distinguish between tiny objects only a fraction of a millimeter or so in size. The width of a human hair is a good example of this viewing limit.

Improvements in optical microscopes over the next three centuries allowed scientists to eventually view objects as tiny as biological cells and bacteria. Nevertheless, optical microscopes are limited in their ultimate resolution by the wavelength of visible light, which is on the order of 0.5 micrometers (ca. 1.64×10^{-6} ft).

To "see" smaller objects, such as viruses, macromolecules, and even individual atoms, scientists needed to develop special new microscopes. The first breakthrough came in the late 1930s, when the German physicist Ernst Ruska (1906–88) invented the electron microscope. The basic principle of this marvelous device involves firing a beam of electrons at the target. Electrons have a much smaller wavelength than visible light, so electron microscopes provide scientists the opportunity of exploring matter from the realm of bacteria (about one micrometer in size) down to nanoscale objects such as small molecules and even individual atoms. Since electron microscope images involve dimensions that are less than the wavelength of visible light, their images normally appear in black and white. Scientists sometimes introduce false color into these high-resolution black-and-white images to assist interpretation of the nanoscale objects by the human eye and brain. Ruska coshared the 1986 Nobel Prize in physics "for his fundamental work in electron optics, and for the design of the first electron microscope."

The transmission electron microscope (TEM) uses electrons traveling through a vacuum as its "light source" to illuminate a target material. Magnets focus the electrons into a tiny beam that impinges on the target material. Some electrons scatter and leave the beam, while others pass through the target and emerge on the other side. These transmitted electrons then produce a shadowlike image of the target material on a fluorescent screen. The image contains various levels of light and darkness—corresponding to the density of the specimen. Scientists record and study these electron microscope images in order to gain new insights into the nanoscale structure of matter.

In 1981, two European researchers working for IBM, Gerd Binning (1947–) and Heinrich Rohrer (1933–), designed the first scanning tunneling microscope (STM). Their original STM was able to examine small samples of matter held in a deeply chilled chamber. At a chamber temperature of 3.6 R (2 K [–271°C]), which is nearly, but not quite, absolute zero, atomic motion—the natural tendency of atoms to move around—slows almost to a halt. Inside the chamber of the STM, the researchers employed an incredibly tiny stylus that was slowly moved over the material object being scanned by a special robotic arm. The stylus was just one atom wide at its tip. Binning and Rohrer had manufactured the device

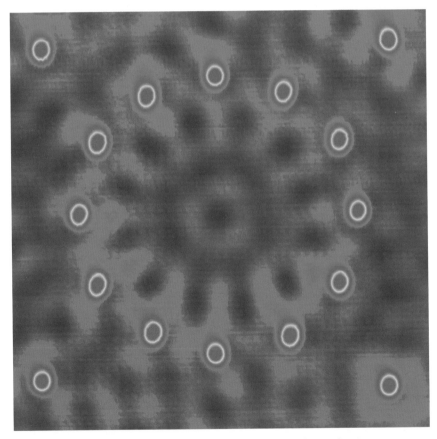

These 12 cobalt atoms arranged in a circle on a surface of copper produce a daisylike pattern from the interference of an electron wave. NIST scientists made this nanoscale image using a unique instrument that, acting autonomously, picks up and places individual atoms anywhere on a surface. *(NIST)*

using newly developed micromachining techniques. By measuring the flow of current in the tip as the stylus approached the surface of the object, the two researchers obtained a precise indication of the distance from the tip to the individual atoms on the surface of the sample object. They used computers to transform these data into an image of the atoms, which was then displayed on a monitor. Other researchers soon discovered that the STM could also be used to push and pull atoms around. These activities represented the first time in the history of technology that humans could manipulate objects on so small a scale. Binning and Rohrer coshared the 1986 Nobel Prize in physics "for their design of the scanning tunneling microscope."

However, the use of the STM was limited to objects, such as metals, that had electrically conductive surfaces. In 1985, Binning, in collaboration with Christoph Gerber and Calvin Quate (1923–   ), invented another device called the atomic force microscope (AFM). This device was constructed in such a way that measurements could be made on non-conductive surfaces. Like the STM, the AFM soon entered scientific and engineering service as a device capable of positioning objects as small as an atom. Today, a variety of very special microscopes and atomic probes support the nanotechnology revolution at research facilities in the United States and around the world. The state-of-the-art equipment found at the National Institute of Standards and Technology (NIST) is just one example.

BUCKYBALLS

Another significant milestone in the development of nanotechnology occurred in about 1985, when the researchers Richard E. Smalley (1943–2005), Robert F. Curl, Jr. (1933–   ), and Sir Harold W. Kroto (1939–   ) discovered an amazing molecule that consisted of 60 linked carbon atoms (C_{60}). Smalley named these distinctive clusters of carbon atoms *buckyballs* in honor of the famous American architect Richard Buckminster Fuller (1895–1983), who promoted the use of geodesic domes. Scientists had known for centuries that carbon consisted of two allotropes, namely diamond and graphite, so their collaborative discovery of another carbon allotrope quite literally rocked the world of chemistry.

Buckyballs are very hard to break apart. When slammed against a solid object or squeezed, they bounce back. The cluster of 60 carbon atoms is especially stable. It has a hollow, icosahedral structure in which the

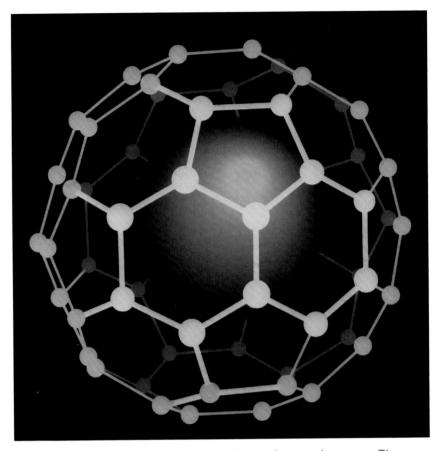

This is a computer model of a buckyball encaging another atom. The buckyball, a 60-atom cluster of carbon atoms (C_{60}), joins diamond and graphite as an allotrope of pure carbon. About one nanometer in diameter, buckyballs are incredibly stable. *(DOE)*

bonds between the carbon atoms resemble the patterns on a soccer ball. Smalley, Curl, and Kroto shared the 1996 Nobel Prize in chemistry for their "discovery of carbon atoms found in the form of a ball."

Identification of this new allotrope of carbon sparked broad interest in the chemistry of an entire family of hollow carbon structures, which scientists now refer to collectively as *fullerenes*. Formed when vaporized carbon condenses in an inert atmosphere, fullerenes include a wide range of shapes and sizes—including nanotubes, which are of interest in electronics and hydrogen storage. Nanotubes are cylindrically shaped tubes of carbon atoms about one nanometer in diameter that are stronger than steel and can conduct electricity. The walls of such nanotubes have the same soccer ball–like structure as do buckyballs, but they come rolled up into long tubes. Since they come in many variations, highly versatile fullerenes could have many potential applications. Researchers think that fullerene structures can eventually be manipulated to produce superconducting salts, new catalysts, new three-dimensional polymers, and biologically active compounds.

THE PROMISE OF NANOTECHNOLOGY

What will the future be like if scientists and engineers develop procedures and devices that can efficiently manipulate matter one atom or molecule at a time? According to most researchers in the field, nanotechnology promises to make major contributions toward solving some of today's most serious problems, such as the control of diseases, handling the adverse aspects of global change, fighting industrial pollution, cleaning up toxic waste sites, and improving food production. There is also the highly anticipated nanoscale revolution in electronics and computer technology.

Perhaps the greatest promise of nanotechnology is to change the way materials and products are produced in the future. Nanostructures, polymers, metals, ceramics, and other materials could be custom designed with greatly improved mechanical, chemical, and physical properties. The ability to build things one atom or molecule at a time suggests the development of entirely new classes of incredible materials. Nanotechnology has the potential to make products lighter, smarter, stronger, cleaner, more highly uniform and precise, and, perhaps best of all, less expensive.

The ability to synthesize nanoscale building blocks, which have precisely controlled dimensions and composition, and then to assemble these designer building blocks into larger structures with unique properties and functions could revolutionize major segments of industry. Scientists suggest that nanostructuring will provide programmable materials. Molecular (or cluster) manufacturing could lead to a host of innovative devices based on new scientific principles and architectures of material assembly at the nanoscale level. Materials scientists speculate that many interesting structures not previously observed in nature will be discovered and harvested. They cite the buckyball as just the beginning.

Nanotechnology promises to transform medicine by providing previously unthinkable tools and procedures for curing some of humankind's most notorious diseases. Imagine what a modern physician could do if he or she were able to perform nanoscale surgery—repairing or replacing individual cells in a patient's body. Nanotechnology advocates postulate that physical diseases can be regarded as the result of misarranged molecules and malfunctioning cells in the human body. Through nanomedicine, physicians might someday be able to reach unobtrusively into the body and effect a cure. Future nanoscale medical devices should allow caregivers to selectively repair mutations in DNA, neutralize cancerous cells, and combat toxic chemicals or viruses that threaten the body. Medical researchers should eventually be able to use nanoscale material manipulation techniques and sensory devices to probe and characterize living cells and networks of healthy molecular structures. Research programs capable of such detailed investigation of living biological materials will result in nothing short of a revolution in the life sciences.

Human lifetimes could be greatly extended as advances in nanotechnology result in the development of more durable, rejection-resistant artificial tissues and organs. Precision therapeutic medicines could be based on the accurate delivery of customized nanoscale pharmaceuticals to individual cells or molecular networks in all parts of the body, including those previously inaccessible. Nanoscale sensor systems (injected into, inhaled by, or ingested by a person) could shift the focus of patient care from disease treatment to early detection and prevention.

In the field of nanoelectronics and information technology, engineers envision the ability to construct incredibly small nanoscale circuits and microprocessors. Smaller circuits run faster, and this, in turn, supports faster computing speeds. Nanotechnology offers the promise of making crystalline materials of ultrapurity and with better thermal conductivity and longer operating life.

There are, of course, enormous technical challenges facing the scientists and engineers who currently labor at the frontiers of nanotechnology. Many social, ethical, political, and economic issues still need to be carefully examined before laboratory discoveries are released to widespread application. Unbiased and careful assessments of potential risks are necessary before human-engineered nanoscale devices are released

(opposite page) Electron microscope images showing a tiny carbon nanotube (upper image) on the hair of an ant's leg (lower image). (Note: Color added for clarity) *(NIST)*

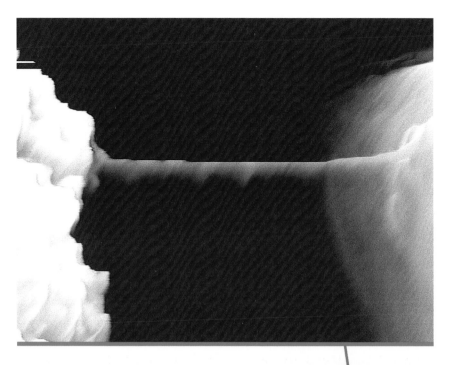

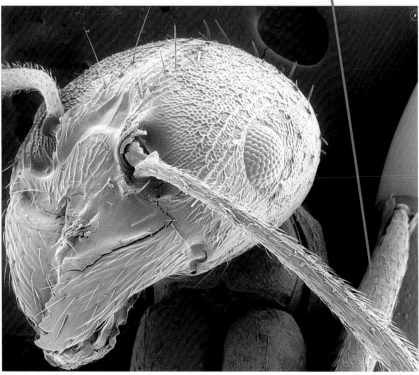

in large quantities into the general environment or the human body. Particular attention must be given to any molecular manufacturing process that involves self-replicating nanoscale systems. Protocols and safeguards must be well established before the application of such systems transitions from laboratory demonstration to large-scale industrial or commercial applications.

Nanotechnology represents an incredibly important plateau in human history, somewhat analogous in its overall potential impact to the discovery and use of fire by humankind's prehistoric ancestors. What is clearly different in this case is the timescale within which the benefits and risks associated with future discoveries can influence the trajectory of civilization. Sweeping technical changes in matter manipulation and materials science could occur in periods as short as decades or possibly even years. Such radical changes in the manipulation of matter will undoubtedly be accompanied by inevitable stresses on existing environmental, social, and economic infrastructures. The price of harvesting huge benefits from this anticipated era of rapid technical progress in materials science is heightened social and ethical vigilance.

CHAPTER 11

Conclusion

Throughout human history, people have wondered about matter. The story of materials science is also the story of civilization. At first, individuals used a variety of interesting but nonscientific approaches to explain the behavior of matter, but in the late 16th century, the true nature of matter and its relationship with energy began to slowly emerge.

As part of the Scientific Revolution that swept across western Europe, inquisitive individuals began quantifying the properties of matter. Assisted by carefully performed and well-documented experiments, they soon discovered a series of interesting relationships. Speculation, philosophical conjecture, and alchemy gave way to the scientific method, with its organized investigation of the material world. Collectively, the story of this magnificent intellectual unfolding represents one of the great cultural legacies in human history—comparable to the control of fire and the invention of the alphabet. The intellectual curiosity and hard work of many pioneering scientists set the human race on a trajectory of discovery, a trajectory that not only enabled today's global civilization but also has opened up the entire universe to understanding and exploration.

As instruments and procedures improved, scientists were able to make even better measurements. Soon matter became more clearly understood on both a macroscopic and microscopic scale. The scientists of the 20th century exponentially accelerated these efforts, and the long-hidden nature of matter emerged on an atomic and even subatomic basis. Nuclear scientists continue to explore nature on its minutest scale, while astrophysicists continue to discover exotic states of matter on the grandest of cosmic scales. Cosmologists and astrophysicists now describe matter as being clumped into enormous clusters and superclusters of galaxies. The quest for these scientists is to explain how the observable universe, consisting of understandable forms of matter and energy, is also immersed in and influenced by mysterious forms of matter and energy called dark matter and dark energy, respectively.

The story of matter and its intimate relationship with human civilization continues today, but at a potentially more rapid pace due to the new global research emphasis on nanotechnology. As an intelligent species suddenly endowed with great powers to manipulate matter and energy down to the atomic scale, people must learn to exercise responsible collective thinking. With great technical power comes great social responsibility—not only for fellow human beings, but for all the other living creatures that share this magnificent world.

Will human manipulation of matter and energy on a very large scale extend beyond Earth in coming centuries? One such notion involves a gigantic astroengineering project consisting of a huge swarm of artificial biospheres encircling the Sun. This concept is the brainchild of the British-American theoretical physicist Freeman John Dyson (1923–). The incredible assembly, sometimes called a *Dyson sphere,* would involve future human beings harvesting the entire material and energy resources of the solar system. For anyone curious about how much effort this would entail: Astronomers estimate that there is about 4.4×10^{27} lbm (2×10^{27} kg) of planetary material available in the entire solar system (excluding the Sun itself). The Sun radiates energy at a rate of about 3.79×10^{23} Btu/s (4×10^{26} J/s) and should continue to do so at this rate for the next 5 billion years. If Dyson's visionary notion takes hold, a few thousand years from now the Sun could be surrounded by a swarm of superbly engineered planetoids containing trillions of people.

For the reader who wishes to speculate about how human beings might manipulate matter and energy even further into the future, imagine a self-replicating robot star probe that would carry a payload of special nanotechnology devices. Sometime in the early 25th century, the human race might elect to send this pioneering self-replicating system out of the solar system on an incredible interstellar voyage. The overall mission of the hypothetical star probe would be to trigger a wave of life in other star systems. This wave (or, more correctly, exponentially growing bubble) of replicating robot systems could eventually reach across the entire Milky Way galaxy—the name of the original self-replicating robot star probe: *Fiat Vita* (Let there be life).

Appendix

Scientists correlate the properties of the elements portrayed in the periodic table with their electron configurations. Since, in a neutral atom, the number of electrons equals the number of protons, they arrange the elements in order of their increasing atomic number (Z). The modern periodic table has seven horizontal rows (called periods) and 18 vertical columns (called groups). The properties of the elements in a particular row vary across it, providing the concept of periodicity.

There are several versions of the periodic table used in modern science. The International Union of Pure and Applied Chemistry (IUPAC) recommends labeling the vertical columns from 1 to 18, starting with hydrogen (H) as the top of group 1 and ending with helium (He) as the top of group 18. The IUPAC further recommends labeling the periods (rows)

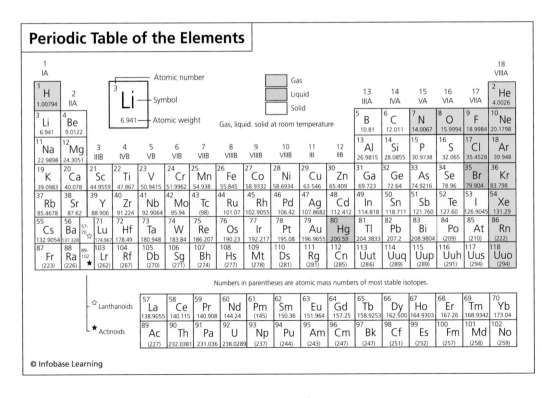

Periodic Table of the Elements

© Infobase Learning

from 1 to 7. Hydrogen (H) and helium (He) are the only two elements found in period (row) 1. Period 7 starts with francium (Fr) and includes the actinide series as well as the transactinides (very short-lived, human-made, super-heavy elements).

The row (or period) in which an element appears in the periodic table tells scientists how many electron shells an atom of that particular element possesses. The column (or group) lets scientists know how many electrons to expect in an element's outermost electron shell. Scientists call an electron residing in an atom's outermost shell a valence electron. Chemists have learned that it is these valence electrons that determine the chemistry of a particular element. The periodic table is structured such that all the elements in the same column (group) have the same number of valence electrons. The elements that appear in a particular column (group) display similar chemistry.

ELEMENTS LISTED BY ATOMIC NUMBER

1	H	Hydrogen	17	Cl	Chlorine
2	He	Helium	18	Ar	Argon
3	Li	Lithium	19	K	Potassium
4	Be	Beryllium	20	Ca	Calcium
5	B	Boron	21	Sc	Scandium
6	C	Carbon	22	Ti	Titanium
7	N	Nitrogen	23	V	Vanadium
8	O	Oxygen	24	Cr	Chromium
9	F	Fluorine	25	Mn	Manganese
10	Ne	Neon	26	Fe	Iron
11	Na	Sodium	27	Co	Cobalt
12	Mg	Magnesium	28	Ni	Nickel
13	Al	Aluminum	29	Cu	Copper
14	Si	Silicon	30	Zn	Zinc
15	P	Phosphorus	31	Ga	Gallium
16	S	Sulfur	32	Ge	Germanium

(continues)

ELEMENTS LISTED BY ATOMIC NUMBER *(continued)*

33	As	Arsenic	62	Sm	Samarium
34	Se	Selenium	63	Eu	Europium
35	Br	Bromine	64	Gd	Gadolinium
36	Kr	Krypton	65	Tb	Terbium
37	Rb	Rubidium	66	Dy	Dysprosium
38	Sr	Strontium	67	Ho	Holmium
39	Y	Yttrium	68	Er	Erbium
40	Zr	Zirconium	69	Tm	Thulium
41	Nb	Niobium	70	Yb	Ytterbium
42	Mo	Molybdenum	71	Lu	Lutetium
43	Tc	Technetium	72	Hf	Hafnium
44	Ru	Ruthenium	73	Ta	Tantalum
45	Rh	Rhodium	74	W	Tungsten
46	Pd	Palladium	75	Re	Rhenium
47	Ag	Silver	76	Os	Osmium
48	Cd	Cadmium	77	Ir	Iridium
49	In	Indium	78	Pt	Platinum
50	Sn	Tin	79	Au	Gold
51	Sb	Antimony	80	Hg	Mercury
52	Te	Tellurium	81	Tl	Thallium
53	I	Iodine	82	Pb	Lead
54	Xe	Xenon	83	Bi	Bismuth
55	Cs	Cesium	84	Po	Polonium
56	Ba	Barium	85	At	Astatine
57	La	Lanthanum	86	Rn	Radon
58	Ce	Cerium	87	Fr	Francium
59	Pr	Praseodymium	88	Ra	Radium
60	Nd	Neodymium	89	Ac	Actinium
61	Pm	Promethium	90	Th	Thorium

91	Pa	Protactinium		105	Db	Dubnium
92	U	Uranium		106	Sg	Seaborgium
93	Np	Neptunium		107	Bh	Bohrium
94	Pu	Plutonium		108	Hs	Hassium
95	Am	Americium		109	Mt	Meitnerium
96	Cm	Curium		110	Ds	Darmstadtium
97	Bk	Berkelium		111	Rg	Roentgenium
98	Cf	Californium		112	Cn	Copernicum
99	Es	Einsteinium		113	Uut	Ununtrium
100	Fm	Fermium		114	Uuq	Ununquadium
101	Md	Mendelevium		115	Uup	Ununpentium
102	No	Nobelium		116	Uuh	Ununhexium
103	Lr	Lawrencium		117	Uus	Ununseptium
104	Rf	Rutherfordium		118	Uuo	Ununoctium

Chronology

Civilization is essentially the story of the human mind understanding and gaining control over matter. The chronology presents some of the major milestones, scientific breakthroughs, and technical developments that formed the modern understanding of matter. Note that dates prior to 1543 are approximate.

13.7 BILLION YEARS AGO . . . . . Big bang event starts the universe.

13.3 BILLION YEARS AGO . . . . . The first stars form and begin to shine intensely.

4.5 BILLION YEARS AGO . . . . . . Earth forms within the primordial solar nebula.

3.6 BILLION YEARS AGO . . . . . . Life (simple microorganisms) appears in Earth's oceans.

2,000,000–100,000 B.C.E. . . . Early hunters of the Lower Paleolithic learn to use simple stone tools, such as handheld axes.

100,000–40,000 B.C.E. . . . . Neanderthal man of Middle Paleolithic lives in caves, controls fire, and uses improved stone tools for hunting.

40,000–10,000 B.C.E. . . . . . During the Upper Paleolithic, Cro-Magnon man displaces Neanderthal man. Cro-Magnon people develop more organized hunting and fishing activities using improved stone tools and weapons.

8000–3500 B.C.E. . . . . Neolithic Revolution takes place in the ancient Middle East as people shift their dependence for subsistence from hunting and gathering to crop cultivation and animal domestication.

3500–1200 B.C.E. . . . . Bronze Age occurs in the ancient Middle East, when metalworking artisans start using bronze (a copper and tin alloy) to make weapons and tools.

1200–600 B.C.E. . . . . People in the ancient Middle East enter the Iron Age. Eventually, the best weapons and tools are made of steel, an alloy of iron and varying amounts of carbon. The improved metal tools and weapons spread to Greece and later to Rome.

1000 B.C.E. .... By this time, people in various ancient civilizations have discovered and are using the following chemical elements (in alphabetical order): carbon (C), copper (Cu), gold (Au), iron (Fe), lead (Pb), mercury (Hg), silver (Ag), sulfur (S), tin (Sn), and zinc (Zn).

650 B.C.E. .... Kingdom of Lydia introduces officially minted gold and silver coins.

600 B.C.E. .... Early Greek philosopher Thales of Miletus postulates that all substances come from water and would eventually turn back into water.

450 B.C.E. .... Greek philosopher Empedocles proposes that all matter is made up of four basic elements (earth, air, water, and fire) that periodically combine and separate under the influence of two opposing forces (love and strife).

430 B.C.E. .... Greek philosopher Democritus proposes that all things consist of changeless, indivisible, tiny pieces of matter called *atoms*.

250 B.C.E. .... Archimedes of Syracuse designs an endless screw, later called the Archimedes screw. People use the fluid-moving device to remove water from the holds of sailing ships and to irrigate arid fields.

300 C.E. .... Greek alchemist Zosimos of Panoplis writes the oldest known work describing alchemy.

850 .... The Chinese use gunpowder for festive fireworks. It is a mixture of sulfur (S), charcoal (C), and potassium nitrate (KNO_3).

1247 .... British monk Roger Bacon writes the formula for gunpowder in his encyclopedic work *Opus Majus*.

1250 .... German theologian and natural philosopher Albertus Magnus isolates the element arsenic (As).

1439 .... Johannes Gutenberg successfully incorporates movable metal type in his mechanical printing press. His revolutionary approach to printing depends on a durable, hard metal alloy called type metal, which consists of a mixture of lead (Pb), tin (Sn), and antimony (Sb).

1543 .... Start of the Scientific Revolution. Polish astronomer Nicholas Copernicus promotes heliocentric (Sun-centered) cosmology with his deathbed publication of *On the Revolutions of Celestial Orbs.*

1638 .... Italian scientist Galileo Galilei publishes extensive work on solid mechanics, including uniform acceleration, free fall, and projectile motion.

1643 .... Italian physicist Evangelista Torricelli designs the first mercury barometer and then records the daily variation of atmospheric pressure.

1661 .... Irish-British scientist Robert Boyle publishes *The Sceptical Chymist,* in which he abandons the four classical Greek elements (earth, air, water, and fire) and questions how alchemists determine what substances are elements.

1665 .... British scientist Robert Hooke publishes *Micrographia,* in which he describes pioneering applications of the optical microscope in chemistry, botany, and other scientific fields.

1667 .... The work of German alchemist Johann Joachim Becher forms the basis of the phlogiston theory of heat.

1669 .... German alchemist Hennig Brand discovers the element phosphorous (P).

1678 .... Robert Hooke studies the action of springs and reports that the extension (or compression) of an elastic material takes place in direct proportion to the force exerted on the material.

1687 .... British physicist Sir Isaac Newton publishes *The Principia.* His work provides the mathematical foundations for understanding (from a classical physics perspective) the motion of almost everything in the physical universe.

1738 .... Swiss mathematician Daniel Bernoulli publishes *Hydrodynamica*. In this seminal work, he identifies the relationships between density, pressure, and velocity in flowing fluids.

1748 .... While conducting experiments with electricity, American statesman and scientist Benjamin Franklin coins the term *battery*.

1754 .... Scottish chemist Joseph Black discovers a new gaseous substance, which he calls "fixed air." Other scientists later identify it as carbon dioxide (CO_2).

1764 .... Scottish engineer James Watt greatly improves the Newcomen steam engine. Watt steam engines power the First Industrial Revolution.

1772 .... Scottish physician and chemist Daniel Rutherford isolates a new colorless gaseous substance, calling it "noxious air." Other scientists soon refer to the new gas as nitrogen (N_2).

1785 .... French scientist Charles-Augustin de Coulomb performs experiments that lead to the important law of electrostatics, later known as Coulomb's law.

1789 .... French chemist Antoine-Laurent Lavoisier publishes *Treatise of Elementary Chemistry*, the first modern textbook on chemistry. Lavoisier also promotes the caloric theory of heat.

1800 .... Italian physicist Count Alessandro Volta invents the voltaic pile. His device is the forerunner of the modern electric battery.

1803 .... British schoolteacher and chemist John Dalton revives the atomic theory of matter. From his experiments, he concludes that all matter consists of combinations of atoms and that all the atoms of a particular element are identical.

1807	British chemist Sir Humphry Davy discovers the element potassium (K) while experimenting with caustic potash (KOH). Potassium is the first metal isolated by the process of electrolysis.
1811	Italian physicist Amedeo Avogadro proposes that equal volumes of different gases under the same conditions of pressure and temperature contain the same number of molecules. Scientists call this important hypothesis Avogadro's law.
1820	Danish physicist Hans Christian Ørsted discovers a relationship between magnetism and electricity.
1824	French military engineer Sadi Carnot publishes *Reflections on the Motive Power of Fire*. Despite the use of caloric theory, his work correctly identifies the general thermodynamic principles that govern the operation and efficiency of all heat engines.
1826	French scientist Andre-Marie Ampere experimentally formulates the relationship between electricity and magnetism.
1827	Experiments performed by German physicist George Simon Ohm indicate a fundamental relationship among voltage, current, and resistance.
1828	Swedish chemist Jöns Jacob Berzelius discovers the element thorium (Th).
1831	British experimental scientist Michael Faraday discovers the principle of electromagnetic induction. This principle is the basis for the electric dynamo.
	Independent of Faraday, the American physicist Joseph Henry publishes a paper describing the electric motor (essentially a reverse dynamo).
1841	German physicist and physician Julius Robert von Mayer states the conservation of energy principle, namely that energy can neither be created nor destroyed.

1847 . . . . British physicist James Prescott Joule experimentally determines the mechanical equivalent of heat. Joule's work is a major step in developing the modern science of thermodynamics.

1866 . . . . Swedish scientist-industrialist Alfred Nobel finds a way to stabilize nitroglycerin and calls the new chemical explosive mixture dynamite.

1869 . . . . Russian chemist Dmitri Mendeleev introduces a periodic listing of the 63 known chemical elements in *Principles of Chemistry.* His periodic table includes gaps for elements predicted but not yet discovered.

American printer John W. Hyatt formulates celluloid, a flammable thermoplastic material made from a mixture of cellulose nitrate, alcohol, and camphor.

1873 . . . . Scottish mathematician and theoretical physicist James Clerk Maxwell publishes *Treatise on Electricity and Magnetism.*

1876 . . . . American physicist and chemist Josiah Willard Gibbs publishes *On the Equilibrium of Heterogeneous Substances.* This compendium forms the theoretical foundation of physical chemistry.

1884 . . . . Swedish chemist Svante Arrhenius proposes that electrolytes split or dissociate into electrically opposite positive and negative ions.

1888 . . . . German physicist Heinrich Rudolf Hertz produces and detects radio waves.

1895 . . . . German physicist Wilhelm Conrad Roentgen discovers X-rays.

1896 . . . . While investigating the properties of uranium salt, French physicist Antoine-Henri Becquerel discovers radioactivity.

1897 . . . . British physicist Sir Joseph John Thomson performs experiments that demonstrate the existence of the electron—the first subatomic particle discovered.

1898	French scientists Pierre and (Polish-born) Marie Curie announce the discovery of two new radioactive elements, polonium (Po) and radium (Ra).
1900	German physicist Max Planck postulates that blackbodies radiate energy only in discrete packets (or quanta) rather than continuously. His hypothesis marks the birth of quantum theory.
1903	New Zealand–born British physicist Baron Ernest Rutherford and British radiochemist Frederick Soddy propose the law of radioactive decay.
1904	German physicist Ludwig Prandtl revolutionizes fluid mechanics by introducing the concept of the boundary layer and its role in fluid flow.
1905	Swiss-German-American physicist Albert Einstein publishes the special theory of relativity, including the famous mass-energy equivalence formula ($E = mc^2$).
1907	Belgian-American chemist Leo Baekeland formulates bakelite. This synthetic thermoplastic material ushers in the age of plastics.
1911..................	Baron Ernest Rutherford proposes the concept of the atomic nucleus based on the startling results of an alpha particle–gold foil scattering experiment.
1912	German physicist Max von Laue discovers that X-rays are diffracted by crystals.
1913	Danish physicist Niels Bohr presents his theoretical model of the hydrogen atom—a brilliant combination of atomic theory with quantum physics. Frederick Soddy proposes the existence of isotopes.
1914	British physicist Henry Moseley measures the characteristic X-ray lines of many chemical elements.
1915	Albert Einstein presents his general theory of relativity, which relates gravity to the curvature of space-time.

1919 . . . . Baron Ernest Rutherford bombards nitrogen (N) nuclei with alpha particles, causing the nitrogen nuclei to transform into oxygen (O) nuclei and to emit protons (hydrogen nuclei).

British physicist Francis Aston uses the newly invented mass spectrograph to identify more than 200 naturally occurring isotopes.

1923 . . . . American physicist Arthur Holly Compton conducts experiments involving X-ray scattering that demonstrate the particle nature of energetic photons.

1924 . . . . French physicist Louis-Victor de Broglie proposes the particle-wave duality of matter.

1926 . . . . Austrian physicist Erwin Schrödinger develops quantum wave mechanics to describe the dual wave-particle nature of matter.

1927 . . . . German physicist Werner Heisenberg introduces his uncertainty principle.

1929 . . . . American astronomer Edwin Hubble announces that his observations of distant galaxies suggest an expanding universe.

1932 . . . . British physicist Sir James Chadwick discovers the neutron.

British physicist Sir John Cockcroft and Irish physicist Ernest Walton use a linear accelerator to bombard lithium (Li) with energetic protons, producing the first artificial disintegration of an atomic nucleus.

American physicist Carl D. Anderson discovers the positron.

1934 . . . . Italian-American physicist Enrico Fermi proposes a theory of beta decay that includes the neutrino. He also starts to bombard uranium with neutrons and discovers the phenomenon of slow neutrons.

1938 .... German chemists Otto Hahn and Fritz Strassmann bombard uranium with neutrons and detect the presence of lighter elements. Austrian physicist Lise Meitner and Austrian-British physicist Otto Frisch review Hahn's work and conclude in early 1939 that the German chemists had split the atomic nucleus, achieving neutron-induced nuclear fission.

E.I. du Pont de Nemours & Company introduces a new thermoplastic material called nylon.

1941 .... American nuclear scientist Glenn T. Seaborg and his associates use the cyclotron at the University of California, Berkeley, to synthesize plutonium (Pu).

1942 .... Modern nuclear age begins when Enrico Fermi's scientific team at the University of Chicago achieves the first self-sustained, neutron-induced fission chain reaction at Chicago Pile One (CP-1), a uranium-fueled, graphite-moderated atomic pile (reactor).

1945 .... American scientists successfully detonate the world's first nuclear explosion, a plutonium-implosion device code-named Trinity.

1947 .... American physicists John Bardeen, Walter Brattain, and William Shockley invent the transistor.

1952 .... A consortium of 11 founding countries establishes CERN, the European Organization for Nuclear Research, at a site near Geneva, Switzerland.

United States tests the world's first thermonuclear device (hydrogen bomb) at the Enewetak Atoll in the Pacific Ocean. Code-named Ivy Mike, the experimental device produces a yield of 10.4 megatons.

1964 .... German-American physicist Arno Allen Penzias and American physicist Robert Woodrow Wilson detect the cosmic microwave background (CMB).

1967 .... German-American physicist Hans Albrecht Bethe receives the 1967 Nobel Prize in physics for his theory of thermonuclear reactions being responsible for energy generation in stars.

1969 .... On July 20, American astronauts Neil Armstrong and Edwin "Buzz" Aldrin successfully land on the Moon as part of NASA's *Apollo 11* mission.

1972 .... NASA launches the *Pioneer 10* spacecraft. It eventually becomes the first human-made object to leave the solar system on an interstellar trajectory

1985 .... American chemists Robert F. Curl, Jr., and Richard E. Smalley, collaborating with British astronomer Sir Harold W. Kroto, discover the buckyball, an allotrope of pure carbon.

1996 .... Scientists at CERN (near Geneva, Switzerland) announce the creation of antihydrogen, the first human-made antimatter atom.

1998 .... Astrophysicists investigating very distant Type 1A supernovae discover that the universe is expanding at an accelerated rate. Scientists coin the term *dark energy* in their efforts to explain what these observations physically imply.

2001 .... American physicist Eric A. Cornell, German physicist Wolfgang Ketterle, and American physicist Carl E. Wieman share the 2001 Nobel Prize in physics for their fundamental studies of the properties of Bose-Einstein condensates.

2005 .... Scientists at the Lawrence Livermore National Laboratory (LLNL) in California and the Joint Institute for Nuclear Research (JINR) in Dubna, Russia, perform collaborative experiments that establish the existence of super-heavy element 118, provisionally called ununoctium (Uuo).

2008 .... An international team of scientists inaugurates the world's most powerful particle accelerator, the Large Hadron Collider (LHC), located at the CERN laboratory near Geneva, Switzerland.

2009 .... British scientist Charles Kao, American scientist Willard Boyle, and American scientist George Smith share the 2009 Nobel Prize in physics for their pioneering efforts in fiber optics and imaging semiconductor devices, developments that unleashed the information technology revolution.

2010 .... Element 112 is officially named Copernicum (Cn) by the IUPAC in honor of Polish astronomer Nicholas Copernicus (1473–1543), who championed heliocentric cosmology.

Scientists at the Joint Institute for Nuclear Research in Dubna, Russia, announce the synthesis of element 117 (ununseptium [Uus]) in early April.

Glossary

absolute zero   the lowest possible temperature; equal to 0 kelvin (K) (−459.67°F, −273.15°C)

acceleration (a)   rate at which the velocity of an object changes with time

accelerator   device for increasing the velocity and energy of charged elementary particles

acid   substance that produces hydrogen ions (H⁺) when dissolved in water

actinoid (formerly actinide)   series of heavy metallic elements beginning with element 89 (actinium) and continuing through element 103 (lawrencium)

activity   measure of the rate at which a material emits nuclear radiations

air   overall mixture of gases that make up Earth's atmosphere

alchemy   mystical blend of sorcery, religion, and prescientific chemistry practiced in many early societies around the world

alloy   solid solution (compound) or homogeneous mixture of two or more elements, at least one of which is an elemental metal

alpha particle (α)   positively charged nuclear particle emitted from the nucleus of certain radioisotopes when they undergo decay; consists of two protons and two neutrons bound together

alternating current (AC)   electric current that changes direction periodically in a circuit

American customary system of units (also American system)   used primarily in the United States; based on the foot (ft), pound-mass (lbm), pound-force (lbf), and second (s). Peculiar to this system is the artificial construct (based on Newton's second law) that one pound-force equals one pound-mass (lbm) at sea level on Earth

ampere (A)   SI unit of electric current

anode   positive electrode in a battery, fuel cell, or electrolytic cell; oxidation occurs at anode

antimatter   matter in which the ordinary nuclear particles are replaced by corresponding antiparticles

Archimedes principle   the fluid mechanics rule that states that the buoyant (upward) force exerted on a solid object immersed in a fluid equals the weight of the fluid displaced by the object

atom   smallest part of an element, indivisible by chemical means; consists of a dense inner core (nucleus) that contains protons and neutrons and a cloud of orbiting electrons

atomic mass   *See* **relative atomic mass**

atomic mass unit (amu)   1/12 mass of carbon's most abundant isotope, namely carbon-12

atomic number (Z)   total number of protons in the nucleus of an atom and its positive charge

atomic weight   the mass of an atom relative to other atoms. *See also* **relative atomic mass**

battery   electrochemical energy storage device that serves as a source of direct current or voltage

becquerel (Bq)   SI unit of radioactivity; one disintegration (or spontaneous nuclear transformation) per second. *Compare with* **curie**

beta particle (β)   elementary particle emitted from the nucleus during radioactive decay; a negatively charged beta particle is identical to an electron

big bang   theory in cosmology concerning the origin of the universe; postulates that about 13.7 billion years ago, an initial singularity experienced a very large explosion that started space and time. Astrophysical observations support this theory and suggest that the universe has been expanding at different rates under the influence of gravity, dark matter, and dark energy

blackbody   perfect emitter and perfect absorber of electromagnetic radiation; radiant energy emitted by a blackbody is a function only of the emitting object's absolute temperature

black hole   incredibly compact, gravitationally collapsed mass from which nothing can escape

boiling point   temperature (at a specified pressure) at which a liquid experiences a change of state into a gas

Bose-Einstein condensate (BEC)   state of matter in which extremely cold atoms attain the same quantum state and behave essentially as a large "super atom"

boson   general name given to any particle with a spin of an integral number (0, 1, 2, etc.) of quantum units of angular momentum. Carrier particles of all interactions are bosons. *See also* **carrier particle**

brass   alloy of copper (Cu) and zinc (Zn)

British thermal unit (Btu)   amount of heat needed to raise the temperature of 1 lbm of water 1°F at normal atmospheric pressure; 1 Btu = 1,055 J = 252 cal

bronze   alloy of copper (Cu) and tin (Sn)

calorie (cal)   quantity of heat; defined as the amount needed to raise one gram of water 1°C at normal atmospheric pressure; 1 cal = 4.1868 J = 0.004 Btu

carbon dioxide (CO_2)   colorless, odorless, noncombustible gas present in Earth's atmosphere

Carnot cycle   ideal reversible thermodynamic cycle for a theoretical heat engine; represents the best possible thermal efficiency of any heat engine operating between two absolute temperatures (T_1 and T_2)

carrier particle   within the standard model, gluons are carrier particles for strong interactions; photons are carrier particles of electromagnetic interactions; and the W and Z bosons are carrier particles for weak interactions. *See also* **standard model**

catalyst   substance that changes the rate of a chemical reaction without being consumed or changed by the reaction

cathode   negative electrode in a battery, fuel cell, electrolytic cell, or electron (discharge) tube through which a primary stream of electrons enters a system

chain reaction   reaction that stimulates its own repetition. *See also* **nuclear chain reaction**

change of state   the change of a substance from one physical state to another; the atoms or molecules are structurally rearranged without experiencing a change in composition. Sometimes called change of phase or phase transition

charged particle   elementary particle that carries a positive or negative electric charge

chemical bond(s)   force(s) that holds atoms together to form stable configurations of molecules

chemical property characteristic of a substance that describes the manner in which the substance will undergo a reaction with another substance, resulting in a change in chemical composition. *Compare with* **physical property**

chemical reaction involves changes in the electron structure surrounding the nucleus of an atom; a dissociation, recombination, or rearrangement of atoms. During a chemical reaction, one or more kinds of matter (called reactants) are transformed into one or several new kinds of matter (called products)

color charge in the standard model, the charge associated with strong interactions. Quarks and gluons have color charge and thus participate in strong interactions. Leptons, photons, W bosons, and Z bosons do not have color charge and consequently do not participate in strong interactions. *See also* **standard model**

combustion chemical reaction (burning or rapid oxidation) between a fuel and oxygen that generates heat and usually light

composite materials human-made materials that combine desirable properties of several materials to achieve an improved substance; includes combinations of metals, ceramics, and plastics with built-in strengthening agents

compound pure substance made up of two or more elements chemically combined in fixed proportions

compressible flow fluid flow in which density changes cannot be neglected

compression condition when an applied external force squeezes the atoms of a material closer together. *Compare* **tension**

concentration for a solution, the quantity of dissolved substance per unit quantity of solvent

condensation change of state process by which a vapor (gas) becomes a liquid. *The opposite of* **evaporation**

conduction (thermal) transport of heat through an object by means of a temperature difference from a region of higher temperature to a region of lower temperature. *Compare with* **convection**

conservation of mass and energy Einstein's special relativity principle stating that energy (E) and mass (m) can neither be created nor destroyed, but are interchangeable in accordance with the equation $E = mc^2$, where c represents the speed of light

convection fundamental form of heat transfer characterized by mass motions within a fluid resulting in the transport and mixing of the properties of that fluid

coulomb (C) SI unit of electric charge; equivalent to quantity of electric charge transported in one second by a current of one ampere

covalent bond the chemical bond created within a molecule when two or more atoms share an electron

creep slow, continuous, permanent deformation of solid material caused by a constant tensile or compressive load that is less than the load necessary for the material to give way (yield) under pressure. *See also* **plastic deformation**

crystal a solid whose atoms are arranged in an orderly manner, forming a distinct, repetitive pattern

curie (Ci) traditional unit of radioactivity equal to 37 billion (37×10^9) disintegrations per second. *Compare with* **becquerel**

current (I) flow of electric charge through a conductor

dark energy a mysterious natural phenomenon or unknown cosmic force thought responsible for the observed acceleration in the rate of expansion of the universe. Astronomical observations suggest dark energy makes up about 72 percent of the universe

dark matter (nonbaryonic matter) exotic form of matter that emits very little or no electromagnetic radiation. It experiences no measurable interaction with ordinary (baryonic) matter but somehow accounts for the observed structure of the universe. It makes up about 23 percent of the content of the universe, while ordinary matter makes up less than 5 percent

density (ρ) mass of a substance per unit volume at a specified temperature

deposition direct transition of a material from the gaseous (vapor) state to the solid state without passing through the liquid phase. *Compare* with **sublimation**

dipole magnet any magnet with one north and one south pole

direct current (DC) electric current that always flows in the same direction through a circuit

elastic deformation   temporary change in size or shape of a solid due to an applied force (stress); when force is removed the solid returns to its original size and shape

elasticity   ability of a body that has been deformed by an applied force to return to its original shape when the force is removed

elastic modulus   a measure of the stiffness of a solid material; defined as the ratio of stress to strain

electricity   flow of energy due to the motion of electric charges; any physical effect that results from the existence of moving or stationary electric charges

electrode   conductor (terminal) at which electricity passes from one medium into another; positive electrode is the *anode;* negative electrode is the *cathode*

electrolyte   a chemical compound that, in an aqueous (water) solution, conducts an electric current

electromagnetic radiation (EMR)   oscillating electric and magnetic fields that propagate at the speed of light. Includes in order of increasing frequency and energy: radio waves, radar waves, infrared (IR) radiation, visible light, ultraviolet radiation, X-rays, and gamma rays

electron (e)   stable elementary particle with a unit negative electric charge (1.602×10^{-19} C). Electrons form an orbiting cloud, or shell, around the positively charged atomic nucleus and determine an atom's chemical properties

electron volt (eV)   energy gained by an electron as it passes through a potential difference of one volt; one electron volt has an energy equivalence of 1.519×10^{-22} Btu = 1.602×10^{-19} J

element   pure chemical substance indivisible into simpler substances by chemical means; all the atoms of an element have the same number of protons in the nucleus and the same number of orbiting electrons, although the number of neutrons in the nucleus may vary

elementary particle   a fundamental constituent of matter; the basic atomic model suggests three elementary particles: the proton, neutron, and electron. *See also* **fundamental particle**

endothermic reaction   chemical reaction requiring an input of energy to take place. *Compare* **exothermic reaction**

energy (E) capacity to do work; appears in many different forms, such as mechanical, thermal, electrical, chemical, and nuclear

entropy (S) measure of disorder within a system; as entropy increases, energy becomes less available to perform useful work

evaporation physical process by which a liquid is transformed into a gas (vapor) at a temperature below the boiling point of the liquid. *Compare with* **sublimation**

excited state state of a molecule, atom, electron, or nucleus when it possesses more than its normal energy. *Compare with* **ground state**

exothermic reaction chemical reaction that releases energy as it takes place. *Compare with* **endothermic reaction**

fatigue weakening or deterioration of metal or other material that occurs under load, especially under repeated cyclic or continued loading

fermion general name scientists give to a particle that is a matter constituent. Fermions are characterized by spin in odd half-integer quantum units (namely, 1/2, 3/2, 5/2, etc.); quarks, leptons, and baryons are all fermions

fission (nuclear) splitting of the nucleus of a heavy atom into two lighter nuclei accompanied by the release of a large amount of energy as well as neutrons, X-rays, and gamma rays

flavor in the standard model, quantum number that distinguishes different types of quarks and leptons. *See also* **quark; lepton**

fluid mechanics scientific discipline that deals with the behavior of fluids (both gases and liquids) at rest (fluid statics) and in motion (fluid dynamics)

foot-pound (force) (ft-lb$_{force}$) unit of work in American customary system of units; 1 ft-lb$_{force}$ = 1.3558 J

force (F) the cause of the acceleration of material objects as measured by the rate of change of momentum produced on a free body. Force is a vector quantity mathematically expressed by Newton's second law of motion: force = mass × acceleration

freezing point the temperature at which a substance experiences a change from the liquid state to the solid state at a specified pressure; at this temperature, the solid and liquid states of a substance can coexist in equilibrium. *Synonymous with* **melting point**

fundamental particle particle with no internal substructure; in the standard model, any of the six types of quarks or six types of leptons and their antiparticles. Scientists postulate that all other particles are made from a combination of quarks and leptons. *See also* **elementary particle**

fusion (nuclear) nuclear reaction in which lighter atomic nuclei join together (fuse) to form a heavier nucleus, liberating a great deal of energy

g acceleration due to gravity at sea level on Earth; approximately 32.2 ft/s² (9.8 m/s²)

gamma ray (γ) high-energy, very short–wavelength photon of electromagnetic radiation emitted by a nucleus during certain nuclear reactions or radioactive decay

gas state of matter characterized as an easily compressible fluid that has neither a constant volume nor a fixed shape; a gas assumes the total size and shape of its container

gravitational lensing bending of light from a distant celestial object by a massive (gravitationally influential) foreground object

ground state state of a nucleus, atom, or molecule at its lowest (normal) energy level

hadron any particle (such as a baryon) that exists within the nucleus of an atom; made up of quarks and gluons, hadrons interact with the strong force

half-life (radiological) time in which half the atoms of a particular radioactive isotope disintegrate to another nuclear form

heat energy transferred by a temperature difference or thermal process. *Compare* **work**

heat capacity (c) amount of heat needed to raise the temperature of an object by one degree

heat engine thermodynamic system that receives energy in the form of heat and that, in the performance of energy transformation on a working fluid, does work. Heat engines function in thermodynamic cycles

hertz (Hz) SI unit of frequency; equal to one cycle per second

high explosive (HE) energetic material that detonates (rather than burns); the rate of advance of the reaction zone into the unreacted material exceeds the velocity of sound in the unreacted material

horsepower (hp) American customary system unit of power; 1 hp = 550 ft-lb$_{force}$/s = 746 W

hydraulic   operated, moved, or affected by liquid used to transmit energy

hydrocarbon   organic compound composed of only carbon and hydrogen atoms

ideal fluid   *See* **perfect fluid**

ideal gas law   important physical principle: $P\,V = n\,R_u\,T$, where P is pressure, V is volume, T is temperature, n is the number of moles of gas, and R_u is the universal gas constant

incompressible flow   fluid flow in which density changes can be neglected. *Compare with* **compressible flow**

inertia   resistance of a body to a change in its state of motion

infrared (IR) radiation   that portion of the electromagnetic (EM) spectrum lying between the optical (visible) and radio wavelengths

International System of units   *See* **SI unit system**

inviscid fluid   perfect fluid that has zero coefficient of viscosity

ion   atom or molecule that has lost or gained one or more electrons, so that the total number of electrons does not equal the number of protons

ionic bond   formed when one atom gives up at least one outer electron to another atom, creating a chemical bond–producing electrical attraction between the atoms

isotope   atoms of the same chemical element but with different numbers of neutrons in their nucleus

joule (J)   basic unit of energy or work in the SI unit system; 1 J = 0.2388 calorie = 0.00095 Btu

kelvin (K)   SI unit of absolute thermodynamic temperature

kinetic energy (KE)   energy due to motion

lepton   fundamental particle of matter that does not participate in strong interactions; in the standard model, the three charged leptons are the electron (e), the muon (μ), and the tau (τ) particle; the three neutral leptons are the electron neutrino (v_e), the muon neutrino (v_μ), and the tau neutrino (v_τ). A corresponding set of antiparticles also exists. *See also* **standard model**

light-year (ly)   distance light travels in one year; 1 ly $\approx 5.88 \times 10^{12}$ miles (9.46 $\times 10^{12}$ km)

liquid   state of matter characterized as a relatively incompressible flowing fluid that maintains an essentially constant volume but assumes the shape of its container

liter (l or L)   SI unit of volume; 1 L = 0.264 gal

magnet   material or device that exhibits magnetic properties capable of causing the attraction or repulsion of another magnet or the attraction of certain ferromagnetic materials such as iron

manufacturing   process of transforming raw material(s) into a finished product, especially in large quantities

mass (m)   property that describes how much material makes up an object and gives rise to an object's inertia

mass number   *See* **relative atomic mass**

mass spectrometer   instrument that measures relative atomic masses and relative abundances of isotopes

material   tangible substance (chemical, biological, or mixed) that goes into the makeup of a physical object

mechanics   branch of physics that deals with the motions of objects

melting point   temperature at which a substance experiences a change from the solid state to the liquid state at a specified pressure; at this temperature, the solid and liquid states of a substance can coexist in equilibrium. *Synonymous with* **freezing point**

metallic bond   chemical bond created as many atoms of a metallic substance share the same electrons

meter (m)   fundamental SI unit of length; 1 meter = 3.281 feet. British spelling *metre*

metric system   *See* **SI unit system**

metrology   science of dimensional measurement; sometimes includes the science of weighing

microwave (radiation)   comparatively short-wavelength electromagnetic (EM) wave in the radio frequency portion of the EM spectrum

mirror matter   *See* **antimatter**

mixture   a combination of two or more substances, each of which retains its own chemical identity

molarity (M)   concentration of a solution expressed as moles of solute per kilogram of solvent

mole (mol)   SI unit of the amount of a substance; defined as the amount of substance that contains as many elementary units as there are atoms in 0.012 kilograms of carbon-12, a quantity known as Avogadro's number (N_A), which has a value of about 6.022×10^{23} molecules/mole

molecule   smallest amount of a substance that retains the chemical properties of the substance; held together by chemical bonds, a molecule can consist of identical atoms or different types of atoms

monomer   substance of relatively low molecular mass; any of the small molecules that are linked together by covalent bonds to form a polymer

natural material   material found in nature, such as wood, stone, gases, and clay

neutrino (ν)   lepton with no electric charge and extremely low (if not zero) mass; three known types of neutrinos are the electron neutrino (ν_e), the muon neutrino (ν_μ), and the tau neutrino (ν_τ). *See also* **lepton**

neutron (n)   an uncharged elementary particle found in the nucleus of all atoms except ordinary hydrogen. Within the standard model, the neutron is a baryon with zero electric charge consisting of two down (d) quarks and one up (u) quark. *See also* **standard model**

newton (N)   The SI unit of force; 1 N = 0.2248 lbf

nuclear chain reaction   occurs when a fissionable nuclide (such as plutonium-239) absorbs a neutron, splits (or fissions), and releases several neutrons along with energy. A fission chain reaction is self-sustaining when (on average) at least one released neutron per fission event survives to create another fission reaction

nuclear energy   energy released by a nuclear reaction (fission or fusion) or by radioactive decay

nuclear radiation   particle and electromagnetic radiation emitted from atomic nuclei as a result of various nuclear processes, such as radioactive decay and fission

nuclear reaction   reaction involving a change in an atomic nucleus, such as fission, fusion, neutron capture, or radioactive decay

nuclear reactor   device in which a fission chain reaction can be initiated, maintained, and controlled

nuclear weapon   precisely engineered device that releases nuclear energy in an explosive manner as a result of nuclear reactions involving fission, fusion, or both

nucleon   constituent of an atomic nucleus; a proton or a neutron

nucleus   (plural: nuclei) small, positively charged central region of an atom that contains essentially all of its mass. All nuclei contain both protons and neutrons except the nucleus of ordinary hydrogen, which consists of a single proton

nuclide   general term applicable to all atomic (isotopic) forms of all the elements; nuclides are distinguished by their atomic number, relative mass number (atomic mass), and energy state

ohm (Ω)   SI unit of electrical resistance

oxidation   chemical reaction in which oxygen combines with another substance, and the substance experiences one of three processes: (1) the gaining of oxygen, (2) the loss of hydrogen, or (3) the loss of electrons. In these reactions, the substance being "oxidized" loses electrons and forms positive ions. *Compare with* **reduction**

oxidation-reduction (redox) reaction   chemical reaction in which electrons are transferred between species or in which atoms change oxidation number

particle   minute constituent of matter, generally one with a measurable mass

pascal (Pa)   SI unit of pressure; 1 Pa = 1 N/m^2 = 0.000145 psi

Pascal's principle   when an enclosed (static) fluid experiences an increase in pressure, the increase is transmitted throughout the fluid; the physical principle behind all hydraulic systems

Pauli exclusion principle   postulate that no two electrons in an atom can occupy the same quantum state at the same time; also applies to protons and neutrons

perfect fluid   hypothesized fluid primarily characterized by a lack of viscosity and usually by incompressibility

perfect gas law   *See* **ideal gas law**

periodic table   list of all the known elements, arranged in rows (periods) in order of increasing atomic numbers and columns (groups) by similar physical and chemical characteristics

phase   one of several different homogeneous materials present in a portion of matter under study; the set of states of a large-scale (macroscopic) physical system having relatively uniform physical properties and chemical composition

phase transition   *See* **change of state**

photon   A unit (or particle) of electromagnetic radiation that carries a quantum (packet) of energy that is characteristic of the particular radiation. Photons travel at the speed of light and have an effective momentum, but no mass or electrical charge. In the standard model, a photon is the carrier particle of electromagnetic radiation

photovoltaic cell   *See* **solar cell**

physical property   characteristic quality of a substance that can be measured or demonstrated without changing the composition or chemical identity of the substance, such as temperature and density. *Compare with* **chemical property**

Planck's constant (h)   fundamental physical constant describing the extent to which quantum mechanical behavior influences nature. Equals the ratio of a photon's energy (E) to its frequency (v), namely: $h = E/v = 6.626 \times 10^{-34}$ J-s (6.282×10^{-37} Btu-s). *See also* **uncertainty principle**

plasma   electrically neutral gaseous mixture of positive and negative ions; called the fourth state of matter

plastic deformation   permanent change in size or shape of a solid due to an applied force (stress)

plasticity   tendency of a loaded body to assume a (deformed) state other than its original state when the load is removed

plastics   synthesized family of organic (mainly hydrocarbon) polymer materials used in nearly every aspect of modern life

pneumatic   operated, moved, or effected by a pressurized gas (typically air) that is used to transmit energy

polymer   very large molecule consisting of a number of smaller molecules linked together repeatedly by covalent bonds, thereby forming long chains

positron (e^+ or β^+)   elementary antimatter particle with the mass of an electron but charged positively

pound-force (lbf)   basic unit of force in the American customary system; 1 lbf = 4.448 N

pound-mass (lbm)   basic unit of mass in the American customary system; 1 lbm = 0.4536 kg

power   rate with respect to time at which work is done or energy is transformed or transferred to another location; 1 hp = 550 ft-lb$_{force}$/s = 746 W

pressure (P)   the normal component of force per unit area exerted by a fluid on a boundary; 1 psi = 6,895 Pa

product   substance produced by or resulting from a chemical reaction

proton (p)   stable elementary particle with a single positive charge. In the the standard model, the proton is a baryon with an electric charge of +1; it consists of two up (u) quarks and one down (d) quark. *See also* **standard model**

quantum mechanics   branch of physics that deals with matter and energy on a very small scale; physical quantities are restricted to discrete values and energy to discrete packets called quanta

quark   fundamental matter particle that experiences strong-force interactions. The six flavors of quarks in order of increasing mass are up (u), down (d), strange (s), charm (c), bottom (b), and top (t)

radiation heat transfer   The transfer of heat by electromagnetic radiation that arises due to the temperature of a body; can takes place in and through a vacuum

radioactive isotope   unstable isotope of an element that decays or disintegrates spontaneously, emitting nuclear radiation; also called radioisotope

radioactivity   spontaneous decay of an unstable atomic nucleus, usually accompanied by the emission of nuclear radiation, such as alpha particles, beta particles, gamma rays, or neutrons

radio frequency (RF)   a frequency at which electromagnetic radiation is useful for communication purposes; specifically, a frequency above 10,000 hertz (Hz) and below 3×10^{11} Hz

rankine (R)   American customary unit of absolute temperature. *See also* **kelvin (K)**

reactant   original substance or initial material in a chemical reaction

reduction   portion of an oxidation-reduction (redox) reaction in which there is a gain of electrons, a gain in hydrogen, or a loss of oxygen. *See also* **oxidation-reduction (redox) reaction**

relative atomic mass (A)   total number of protons and neutrons (nucleons) in the nucleus of an atom. Previously called *atomic mass* or *atomic mass number*. *See also* **atomic mass unit**

residual electromagnetic effect   force between electrically neutral atoms that leads to the formation of molecules

residual strong interaction   interaction responsible for the nuclear binding force—that is, the strong force holding hadrons (protons and neutrons) together in the atomic nucleus. *See also* **strong force**

resilience   property of a material that enables it to return to its original shape and size after deformation

resistance (R)   the ratio of the voltage (V) across a conductor to the electric current (I) flowing through it

scientific notation   A method of expressing powers of 10 that greatly simplifies writing large numbers; for example, $3 \times 10^6 = 3,000,000$

SI unit system   international system of units (the metric system), based upon the meter (m), kilogram (kg), and second (s) as the fundamental units of length, mass, and time, respectively

solar cell (photovoltaic cell)   a semiconductor direct energy conversion device that transforms sunlight into electric energy

solid   state of matter characterized by a three-dimensional regularity of structure; a solid is relatively incompressible, maintains a fixed volume, and has a definitive shape

solution   When scientists dissolve a substance in a pure liquid, they refer to the dissolved substance as the *solute* and the host pure liquid as the *solvent*. They call the resulting intimate mixture the solution

spectroscopy   study of spectral lines from various atoms and molecules; emission spectroscopy infers the material composition of the objects that emitted the light; absorption spectroscopy infers the composition of the intervening medium

speed of light *(c)*   speed at which electromagnetic radiation moves through a vacuum; regarded as a universal constant equal to 186,283.397 mi/s (299,792.458 km/s)

stable isotope   isotope that does not undergo radioactive decay

standard model   contemporary theory of matter, consisting of 12 fundamental particles (six quarks and six leptons), their respective antiparticles, and four force carriers (gluons, photons, W bosons, and Z bosons)

state of matter   form of matter having physical properties that are quantitatively and qualitatively different from other states of matter; the three more common states on Earth are solid, liquid, and gas

steady state   condition of a physical system in which parameters of importance (fluid velocity, temperature, pressure, etc.) do not vary significantly with time

strain   the change in the shape or dimensions (volume) of an object due to applied forces; longitudinal, volume, and shear are the three basic types of strain

stress   applied force per unit area that causes an object to deform (experience strain); the three basic types of stress are compressive (or tensile) stress, hydrostatic pressure, and shear stress

string theory   theory of quantum gravity that incorporates Einstein's general relativity with quantum mechanics in an effort to explain space-time phenomena on the smallest imaginable scales; vibrations of incredibly tiny stringlike structures form quarks and leptons

strong force   In the standard model, the fundamental force between quarks and gluons that makes them combine to form hadrons, such as protons and neutrons; also holds hadrons together in a nucleus. *See also* **standard model**

subatomic particle   any particle that is small compared to the size of an atom

sublimation   direct transition of a material from the solid state to the gaseous (vapor) state without passing through the liquid phase. *Compare* **deposition**

superconductivity   the ability of a material to conduct electricity without resistance at a temperature above absolute zero

temperature (T)   thermodynamic property that serves as a macroscopic measure of atomic and molecular motions within a substance; heat naturally flows from regions of higher temperature to regions of lower temperature

tension   condition when applied external forces pull atoms of a material farther apart. *Compare* **compression**

thermal conductivity (k)   intrinsic physical property of a substance; a material's ability to conduct heat as a consequence of molecular motion

thermodynamics   branch of science that treats the relationships between heat and energy, especially mechanical energy

thermodynamic system   collection of matter and space with boundaries defined in such a way that energy transfer (as work and heat) from and to the system across these boundaries can be easily identified and analyzed

thermometer instrument or device for measuring temperature

toughness ability of a material (especially a metal) to absorb energy and deform plastically before fracturing

transmutation transformation of one chemical element into a different chemical element by a nuclear reaction or series of reactions

transuranic element (isotope) human-made element (isotope) beyond uranium on the periodic table

ultraviolet (UV) radiation portion of the electromagnetic spectrum that lies between visible light and X-rays

uncertainty principle Heisenberg's postulate that places quantum-level limits on how accurately a particle's momentum (p) and position (x) can be simultaneously measured. Planck's constant (h) expresses this uncertainty as $\Delta x \times \Delta p \geq h/2\pi$

U.S. customary system of units *See* **American customary system of units**

vacuum relative term used to indicate the absence of gas or a region in which there is a very low gas pressure

valence electron electron in the outermost shell of an atom

van der Waals force generally weak interatomic or intermolecular force caused by polarization of electrically neutral atoms or molecules

vapor gaseous state of a substance

velocity vector quantity describing the rate of change of position; expressed as length per unit of time

velocity of light (c) *See* **speed of light**

viscosity measure of the internal friction or flow resistance of a fluid when it is subjected to shear stress

volatile solid or liquid material that easily vaporizes; volatile material has a relatively high vapor pressure at normal temperatures

volt (V) SI unit of electric potential difference

volume (V) space occupied by a solid object or a mass of fluid (liquid or confined gas)

watt (W) SI unit of power (work per unit time); $1\ W = 1\ J/s = 0.00134\ hp = 0.737\ ft\text{-}lb_{force}/s$

wavelength (λ) the mean distance between two adjacent maxima (or minima) of a wave

weak force   fundamental force of nature responsible for various types of radioactive decay

weight (w)   the force of gravity on a body; on Earth, product of the mass (m) of a body times the acceleration of gravity (g), namely $w = m \times g$

work (W)   energy expended by a force acting though a distance. *Compare* **heat**

X-ray   penetrating form of electromagnetic (EM) radiation that occurs on the EM spectrum between ultraviolet radiation and gamma rays

Further Resources

BOOKS

Allcock, Harry R. *Introduction to Materials Chemistry.* New York: John Wiley & Sons, 2008. A college-level textbook that provides a basic treatment of the principles of chemistry upon which materials science depends.

Angelo, Joseph A., Jr. *Nuclear Technology.* Westport, Conn.: Greenwood Press, 2004. The book provides a detailed discussion of both military and civilian nuclear technology and includes impacts, issues, and future advances.

———. *Encyclopedia of Space and Astronomy.* New York: Facts On File, 2006. Provides a comprehensive treatment of major concepts in astronomy, astrophysics, planetary science, cosmology, and space technology.

Ball, Philip. *Designing the Molecular World: Chemistry at the Frontier.* Princeton, N.J.: Princeton University Press, 1996. Discusses many recent advances in modern chemistry, including nanotechnology and superconductor materials.

———. *Made to Measure: New Materials for the 21st Century.* Princeton, N.J.: Princeton University Press, 1998. Discusses how advanced new materials can significantly influence life in the 21st century.

Bensaude-Vincent, Bernadette, and Isabelle Stengers. *A History of Chemistry.* Cambridge, Mass.: Harvard University Press, 1996. Describes how chemistry emerged as a science and its impact on civilization.

Callister, William D., Jr. *Materials Science and Engineering: An Introduction.* 8th ed. New York: John Wiley & Sons, 2010. Intended primarily for engineers, technically knowledgeable readers will also benefit from this book's introductory treatment of metals, ceramics, polymers, and composite materials.

Charap, John M. *Explaining the Universe: The New Age of Physics.* Princeton, N.J.: Princeton University Press, 2004. Discusses the important discoveries in physics during the 20th century that are influencing civilization.

Close, Frank, et al. *The Particle Odyssey: A Journey to the Heart of the Matter.* New York: Oxford University Press, 2002. A well-illustrated and enjoyable tour of the subatomic world.

Cobb, Cathy, and Harold Goldwhite. *Creations of Fire: Chemistry's Lively History from Alchemy to the Atomic Age.* New York: Plenum Press, 1995. Uses historic circumstances and interesting individuals to describe the emergence of chemistry as a scientific discipline.

Feynman, Richard P. *QED: The Strange Theory of Light and Matter.* Princeton, N.J.: Princeton University Press, 2006. Written by an American Nobel laureate, addresses several key topics in modern physics.

Gordon, J. E. *The New Science of Strong Materials or Why You Don't Fall Through the Floor.* Princeton, N.J.: Princeton University Press, 2006. Discusses the science of structural materials in a manner suitable for both technical and lay audiences.

Hill, John W., and Doris K. Kolb. *Chemistry for Changing Times.* 11th ed. Upper Saddle River, N.J.: Pearson Prentice Hall, 2007. Readable college-level textbook that introduces all the basic areas of modern chemistry.

Krebs, Robert E. *The History and Use of Our Earth's Chemical Elements: A Reference Guide.* 2nd ed. Westport, Conn.: Greenwood Press, 2006. Provides a concise treatment of each of the chemical elements.

Levere, Trevor H. *Transforming Matter: A History of Chemistry from Alchemy to the Buckyball.* Baltimore: Johns Hopkins University Press, 2001. Provides an understandable overview of the chemical sciences from the early alchemists through modern times.

Lutgens, Frederick K., and Edward J. Tarbuck. *The Atmosphere: An Introduction to Meteorology.* 10th ed. Upper Saddle River, N.J.: Pearson Prentice Hall, 2007. Readable college-level textbook that discusses the atmosphere, meteorology, climate, and the physical properties of air.

Mackintosh, Ray, et al. *Nucleus: A Trip into the Heart of Matter.* Baltimore: Johns Hopkins University Press, 2001. Provides a technical though readable explanation of how modern scientists developed their current understanding of the atomic nucleus and the standard model.

Nicolaou, K. C., and Tamsyn Montagnon. *Molecules that Changed the World.* New York: John Wiley & Sons, 2008. Provides an interesting treatment of such important molecules as aspirin, camphor, glucose, quinine, and morphine.

Scerri, Eric R. *The Periodic Table: Its Story and Its Significance.* New York: Oxford University Press, 2007. Provides a detailed look at the periodic table and its iconic role in the practice of modern science.

Smith, William F., and Javad Hashemi. *Foundations of Materials Science and Engineering.* 5th ed. New York: McGraw-Hill, 2006. Provides scientists and engineers of all disciplines an introduction to materials science, including metals, ceramics, polymers, and composite materials. Technically knowledgeable laypersons will find the treatment of specific topics such as biological materials useful.

Strathern, Paul. *Mendeleyev's Dream: The Quest for the Elements.* New York: St. Martin's Press, 2001. Describes the intriguing history of chemistry from the early Greek philosophers to the 19th-century Russian chemist Dmitri Mendeleyev.

Thrower, Peter, and Thomas Mason. *Materials in Today's World.* 3rd ed. New York: McGraw-Hill Companies, 2007. Provides a readable introductory treatment of modern materials science, including biomaterials and nanomaterials.

Trefil, James, and Robert M. Hazen. *Physics Matters: An Introduction to Conceptual Physics.* New York: John Wiley & Sons, 2004. Highly-readable introductory college-level textbook that provides a good overview of physics from classical mechanics to relativity and cosmology. Laypersons will find the treatment of specific topics useful and comprehendible.

Zee, Anthony. *Quantum Field Theory in a Nutshell.* Princeton, N.J.: Princeton University Press, 2003. A reader-friendly treatment of the generally complex and profound physical concepts that constitute quantum field theory.

WEB SITES

To help enrich the content of this book and to make your investigation of matter more enjoyable, the following is a selective list of recommended Web sites. Many of the sites below will also lead to other interesting science-related locations on the Internet. Some sites provide unusual science learning opportunities (such as laboratory simulations) or in-depth educational resources.

American Chemical Society (ACS) is a congressionally chartered independent membership organization that represents professionals at all degree levels and in all fields of science involving chemistry. The ACS Web site includes educational resources for high school and college students. Available online. URL: http://portal.acs.org/portal/acs/corg/content. Accessed on February 12, 2010.

American Institute of Physics (AIP) is a not-for-profit corporation that promotes the advancement and diffusion of the knowledge of physics and its applications to human welfare. This Web site offers an enormous quantity of fascinating information about the history of physics from ancient Greece up to the present day. Available online. URL: http://www.aip.org/aip/. Accessed on February 12, 2010.

Chandra X-ray Observatory (CXO) is a space-based NASA astronomical observatory that observes the universe in the X-ray portion of the electromagnetic spectrum. This Web site contains contemporary information and educational materials about astronomy, astrophysics, and cosmology, including topics such as black holes, neutron stars, dark matter, and dark energy. Available online. URL: http://www.chandra.harvard.edu/. Accessed on February 12, 2010.

The ChemCollective is an online resource for learning about chemistry. Through simulations developed by the Department of Chemistry of Carnegie Mellon University (with funding from the National Science Foundation), a person gets the chance to safely mix chemicals without worrying about accidentally spilling them. Available online. URL: http://www.chemcollective.org/vlab/vlab.php. Accessed on February 12, 2010.

Chemical Heritage Foundation (CHF) maintains a rich and informative collection of materials that describe the history and heritage of the chemical and molecular sciences, technologies, and industries. Available online. URL: http://www.chemheritage.org/. Accessed on February 12, 2010.

Department of Defense (DOD) is responsible for maintaining armed forces of sufficient strength and technology to protect the United States and its citizens from all credible foreign threats. This Web site serves as an efficient access point to activities within the DOD, including those taking place within each of the individual armed services: the U.S. Army, U.S. Navy, U.S. Air Force, and U.S. Marines. As part of national security, the DOD sponsors a large amount of research and development, including activities in materials science, chemistry, physics, and nanotechnology. Available online. URL: http://www.defenselink.mil/. Accessed on February 12, 2010.

Department of Energy (DOE) is the single largest supporter of basic research in the physical sciences in the federal government of the United States. Topics found on this Web site include materials sciences, nanotechnology, energy sciences, chemical science, high-energy physics, and nuclear physics. The Web site also includes convenient links to all of the DOE's national laboratories. Available online. URL: http://energy.gov/. Accessed on February 12, 2010.

Fermi National Accelerator Laboratory (Fermilab) performs research that advances the understanding of the fundamental nature of matter and energy. Fermilab's Web site contains contemporary information about particle physics, the standard model, and the impact of particle physics on society. Available online. URL: http://www.fnal.gov/. Accessed on February 12, 2010.

Hubble Space Telescope (HST) is a space-based NASA observatory that has examined the universe in the (mainly) visible portion of the electromagnetic spectrum. This Web site contains contemporary information and educational materials about astronomy, astrophysics, and cosmology, including topics such as black holes, neutron stars, dark matter, and dark energy. Available online. URL: http://hubblesite.org/. Accessed on February 12, 2010.

Institute and Museum of the History of Science in Florence, Italy, offers a special collection of scientific instruments (some viewable online), including those used by Galileo Galilei. Available online. URL: http://www.imss.fi.it/. Accessed on February 12, 2010.

International Union of Pure and Applied Chemistry (IUPAC) is an international nongovernmental organization that fosters worldwide communications in the chemical sciences and in providing a common language for chemistry that unifies the industrial, academic, and public sectors. Available online. URL: http://www.iupac.org/. Accessed on February 12, 2010.

National Aeronautics and Space Administration (NASA) is the civilian space agency of the U.S. government and was created in 1958 by an act of Congress. NASA's overall mission is to direct, plan, and conduct American civilian (including scientific) aeronautical and space activities for peaceful purposes. Available online. URL: http://www.nasa.gov/. Accessed on February 12, 2010.

National Institute of Standards and Technology (NIST) is an agency of the U.S. Department of Commerce that was founded in 1901 as the nation's first federal physical science research laboratory. The NIST Web site includes contemporary information about many areas of science and engineering, including analytical chemistry, atomic and molecular physics, biometrics, chemical and crystal structure, chemical kinetics, chemistry, construction, environmental data, fire, fluids, material properties, physics, and thermodynamics. Available online. URL: http://www.nist.gov/index.html. Accessed on February 12, 2010.

National Oceanic and Atmospheric Administration (NOAA) was established in 1970 as an agency within the U.S. Department of Commerce to ensure the safety of the general public from atmospheric phenomena and to provide the public with an understanding of Earth's environment and resources. Available online. URL: http://www.noaa.gov/. Accessed on February 12, 2010.

NEWTON: Ask a Scientist is an electronic community for science, math, and computer science educators and students sponsored by the Argonne National Laboratory (ANL) and the U.S. Department of Energy's Office of Science Education. This Web site provides access to a fascinating list of questions and answers involving the following disciplines/topics: astronomy, biology, botany, chemistry, computer science, Earth science, engineering, environmental science, general science, materials science, mathematics, molecular biology, physics, veterinary, weather, and zoology. Available online. URL: http://www.newton.dep.anl.gov/archive.htm. Accessed on February 12, 2010.

Nobel Prizes in Chemistry and Physics. This Web site contains an enormous amount of information about all the Nobel Prizes awarded in physics and chemistry, as well as complementary technical information. Available online. URL: http://nobelprize.org/. Accessed on February 12, 2010.

Periodic Table of Elements. An informative online periodic table of the elements maintained by the Chemistry Division of the Department of Energy's Los Alamos National Laboratory (LANL). Available online. URL: http://periodic.lanl.gov/. Accessed on February 12, 2010.

PhET Interactive Simulations is an ongoing effort by the University of Colorado at Boulder (under National Science Foundation sponsorship) to provide a comprehensive collection of simulations to enhance science learning. The major science categories include physics, chemistry, Earth science, and biology. Available online. URL: http://phet.colorado.edu/index.php. Accessed on February 12, 2010.

ScienceNews is the online version of the magazine of the Society for Science and the Public. Provides insights into the latest scientific achievements and discoveries. Especially useful are the categories Atom and Cosmos, Environment, Matter and Energy, Molecules, and Science and Society. Available online. URL: http://www.sciencenews.org/. Accessed on February 12, 2010.

The Society on Social Implications of Technology (SSIT) of the Institute of Electrical and Electronics Engineers (IEEE) deals with such issues as the environmental, health, and safety implications of technology; engineering ethics; and the social issues related to telecommunications, information technology, and energy. Available online. URL: http://www.ieeessit.org/. Accessed on February 12, 2010.

Spitzer Space Telescope (SST) is a space-based NASA astronomical observatory that observes the universe in the infrared portion of the electromagnetic spectrum. This Web site contains contemporary information and educational materials about astronomy, astrophysics, and cosmology, including the infrared universe, star and planet formation, and infrared radiation. Available online. URL: http://www.spitzer.caltech.edu/. Accessed on February 12, 2010.

Thomas Jefferson National Accelerator Facility (Jefferson Lab) is a U.S. Department of Energy–sponsored laboratory that conducts basic research on the atomic nucleus at the quark level. The Web site includes basic information about the periodic table, particle physics, and quarks. Available online. URL: http://www.jlab.org/. Accessed on February 12, 2010.

United States Geological Survey (USGS) is the agency within the U.S. Department of the Interior that serves the nation by providing reliable scientific information needed to describe and understand Earth, minimize the loss of life and property from natural disasters, and manage water, biological, energy, and mineral resources. The USGS Web site is rich in science information, including the atmosphere and climate,

Earth characteristics, ecology and environment, natural hazards, natural resources, oceans and coastlines, environmental issues, geologic processes, hydrologic processes, and water resources. Available online. URL: http://www.usgs.gov/. Accessed on February 12, 2010.

Index

Italic page numbers indicate illustrations.